A CONCISE NATURAL HISTORY
OF EAST AND WEST FLORIDA

A CONCISE NATURAL HISTORY

OF

Eaſt and *Weſt* FLORIDA;

CONTAINING

An Account of the natural Produce of all the Southern Part of BRITISH AMERICA, in the three Kingdoms of Nature, particularly the Animal and Vegetable.

LIKEWISE,

The artificial Produce now raiſed, or poſſible to be raiſed, and manufactured there, with ſome commercial and political Obſervations in that part of the world; and a chorographical Account of the ſame.

To which is added, by Way of Appendix,

Plain and eaſy Directions to Navigators over the Bank of Bahama, the Coaſt of the two Floridas, the North of Cuba, and the dangerous Gulph Paſſage. Noting alſo, the hitherto unknown watering Places in that Part of America, intended principally for the Uſe of ſuch Veſſels as may be ſo unfortunate as to be diſtreſſed by Weather in that difficult Part of the World.

By Captain *BERNARD ROMANS.*

Illuſtrated with twelve COPPER PLATES, And Two whole Sheet MAPS.

VOL. I.

NEW-YORK
Printed for the AUTHOR, M,DCC,LXXV.

Reproduction of Original Title Page

Reprinted by

A FIREBIRD PRESS BOOK

PELICAN PUBLISHING COMPANY

ISBN 1-56554-613-X

Manufactured in the United States of America
Published by Pelican Publishing Company, Inc.
1000 Burmaster Street, Gretna, Louisiana 70053

PUBLISHER'S NOTE

In reprinting "A Concise Natural History of East and West Florida" by Bernard Romans, no change in the original text has been made except to use the modern "s" in place of the old-fashioned "s" that looks like an "f"; and putting the first person "i" in capitals.

The original edition printed in 1775 by a New York printer is replete with errors of spelling, punctuation, broken parentheses and sometimes grammar. For example, on Page 247 *Boca Ratones* is spelled in one place with one "t" and on the same page, further down, the same word is spelled with two "t" 's. There are various spellings of "Mississippi", sometimes three "s" 's only and sometimes only one "p". There are hundreds of errors of punctuation and much misspelling, even allowing for eighteenth century orthography.

It was thought best not to attempt to correct the mistakes of the original printer and the text has been reprinted, errors and all, exactly as it is in its first edition. The only exceptions are about 20 errors included in an errata slip in the original edition which errors are corrected here.

No attempt was made to reproduce the original book page for page by photographing or lithographing the pages of the original, as this would not have obviated the difficulties of the old-fashioned "s", and the book would have been harder to read. Great care, however, has been exercised in proof reading in an endeavor to keep from the new book any errors that the modern printer may have made in re-setting the type.

The illustrations, with the exception of the frontispiece, which is reproduced in half-tone, are made in line as they make a better reproduction of the originals than if reproduced as half-tone cuts. The illustrations have been checked against the two copies of this rare book in possession of the Library of Congress.

This reprint of Romans' "A Concise Natural History of East and West Florida" is made from a copy of the micro-film of this book owned by the Henry E. Huntington Library and Art Gallery of San Marino, California. We are indebted to the Huntington Library for permission to use their micro-film and thank them for this courtesy.

PELICAN PUBLISHING COMPANY
NEW ORLEANS, LOUISIANA
October 1st, 1961

tion, more familiar to me, which is that of one Mr. Francois, who lives now about five miles below the river Poule: In September 1771, i called there, the old man told me he was then past eighty three years of age, that the old woman, whom i faw putting bread into the oven, was his mother; and that ſhe was one of the firſt women that came from France to this country; i ſaw her about her domeſtick buſineſs in many ways; in a very cheerful manner, ſinging and running from place to place as briſkly as a girl of twenty; Mr. François told me, that at the age of ſixty he fell out of a pine tree, above fifty feet high, with his loins over a fallen one, that he with difficulty recovered, and that had it not been for that accident, he would not, as he thinks, yet have been ſenſible of the heavy hand of time; that he was ſtill a hearty cheerful old man, was evidently to be ſeen; when i came to the river Poule in October 1772, i met the ſame old gentleman fiſhing at the mouth of the river, on my aſking him whether this diverſion was agreeable to him, he told me; that his mother had an inclination to eat fiſh, and he was come to get her a meſs; he was then on foot and had five miles to come to this place, and as much back with his prey, after catching it; a very dutiful ſon this at eighty five! He lives comfortably at an agreeable place, and on the produce of a midling large ſtock of cattle.

Many more of this kind might be mentioned, but theſe two being more univerſally known, i choſe to relate them only. Far otherwiſe was it with our ſons of incontinence, who upon their arrival, and after their firſt taking poſſeſſion of this country, lived there ſo faſt, that their race was

too

Reproduction of Page 12 of the original edition of Romans' "A Concise Natural History of East and West Florida", showing his use of the lower case "i" and the old fashioned "s". There are 39 such usages on this one page.

LIST OF ILLUSTRATIONS

Page

Symbolical Design .... Frontispiece

Swamp Plant — Avena Aquatica Sylvestris 20

Head of a Chicasaw Indian 39

Busts of Chactaw Indians 54

Indian Funeral Platform or "Stage" 60

Head of Creek War Chief 64

Indian Hierogliphics 70

Table of Exports from the Province of Georgia 74-75

Map of Tampa Bay (Chart with soundings) 283

Map of Pensacola Harbour (with soundings) 285

Map of Mobile Bar (with soundings) 287

INTRODUCTION TO NEW EDITION

Beyond the fact that Bernard Romans was born in Holland about 1720 and was educated in England, comparatively little is known of his life before 1755, when he was sent to America by the British government as a civil engineer. For a part of the time between 1760 and 1770 he was the deputy surveyor of Georgia; then he was in East Florida as a surveyor of Lord Egmont's estates on Amelia Island and the St. Johns River. During much of this time he lived in or near St. Augustine, and through his work was able to make many observations in that part of the peninsula. In 1769-1770 he was appointed chief deputy surveyor for the Southern District and first commander of the vessels in service. Later this position enabled him to go, at his own expense, on a lengthy voyage which took him to the Bahamas and the West coast of Florida as far as Pensacola, where he arrived in 1771. Here he was employed by Governor Peter Chester and John Stuart, superintendent of Indian affairs, to survey the extensive area of West Florida and to make maps of this section. When Governor Chester discovered that Romans understood botany, he obtained permission from England for him "to make botanical discoveries." Thus Romans became the king's botanist in the province.

Historians describe Romans as one of the remarkable men who helped build up this country in colonial times. He has been called a universal genius, distinguished not only as surveyor and botanist but as engineer, cartographer, mathematician, writer, seaman, soldier, patriot. He is also known as linguist, artist, and engraver.

Romans the cartographer is equally as important as Romans the writer. His great and extensive map of the Floridas is one of the finest pieces of cartography of the region and is rightly considered the second most important, if not the most important, map of Florida. It was engraved by Paul Revere, according to the account-books of that master engraver. Until comparatively recent years the map was so rare that bibliographers doubted its existence. It is still scarce and a collector's item.

From the title page of the book and from Romans' advertisements of the volume it is certain that the HISTORY and the map were to form a complete work. The title page reads "illustrated with twelve copper plates and two whole sheet maps". Advertisements state: "There is added to the Maps a Book of 500 pages. The Inland Country is very minutely described, and the Maps will explain even the Land laid out on the river Mississippi". However, judging by the way in which subscribers are listed on the introductory pages I-VIII of the book, one could choose to subscribe to both book and map or to the map only.

Both East and West Florida are shown on the map, although the title, "Part of the Province of East Florida", would seem to exclude the western area. The information on the map shows, among other things, rivers and their tributaries, islands, bays, and creeks. Descriptions are given of the coast line, soundings, and the character of the ocean bottom and adjacent lands. Many place names and notes are also given, as along the Mississippi River such places as "English Reach", "Ruins of Fort Bute", "Baton Rouge", and "Fine Plantations of Both Sides".

Not the least interesting feature of the map are the cartouches on each of the two sheets. The dedication of the first sheet is to the Marine Society of the City of New York, and is in a cartouche formed by the seal of the Marine Society, and two symbolical figures, one representing war, the other civilization enlightening savagery. The second sheet has two dedications in cartouches, one "To all Commanders of vessels round the Globe", the other "To the Hon. the Planters in Jamaica and all Merchants Concerned in the trade of that Island". Even though quite elaborate the cartouches are somewhat crudely made. Each dedication is inscribed by B. Romans.

In the lengthy advertisement in the BOSTON GAZETTE for January 10, 1774, Romans stresses the value of his work to traders and to "Merchants who trade to Jamaica, Hispaniola, and the Two Floridas". He gives many comments on and details of book and map, as "In this exceeding difficult and dangerous Navigation are a great many watering Places not hitherto known, which will be described and directed to, and by which Means it is hoped many a poor distressed Crew will be saved from Ruin, even when they perhaps despair of Life". Another lure to subscribers suggests that "The elegance of the map added to its large size, of twelve Feet by seven, will likewise render it an Ornamental Piece of Furniture".

In his writing he may well be considered a "pioneer describer of Florida." Among the very full accounts by contemporaries dealing with the period of the English occupation of Florida—de Brahm, John and Willian Bartram, William Stork—Romans' *A Concise History of East and West Florida,* 1775, is of the first importance, basic to any study of the period. It is also the rarest.

The *History* shows originality in both style and content. According to the late Mr. Phillips of the Library of Congress "the variety of natural, aboriginal, historic, and miscellaneous information which it graphically gives is far more original than a great many pioneer histories, as Filson's *Kentucky,* Cutler's *Ohio,* and Daniel Smith's *Tennessee."* And his material is so full and so accurate that modern historians have found little to add or to change.

The natural features of Florida are minutely described. He shows clearly that this was a rich country which could produce rare and valuable crops. Typical products of the colony were excellent indigo, wild pulse for feeding cattle, and timber which is described as unsurpassed in quantity, quality, and variety. As to fish, he gives a distinct impression that the fishing industry in West Florida makes up in quality what it lacks in quantity.

He is copious in describing the manners and customs of the native tribes. He observes and carefully portrays the Chickasaw, Choctaw, and Creek Indians: their physical appearance, manners, customs, dress, habits, their similarities and their differences. With all their bad traits Romans finds that they have at least one outstanding virtue —hospitality.

He is naturally interested in the English provinces, centering his description more upon the East coast and especially St. Augustine. He is sometimes critical in his attitude. He objects strongly to the barracks and fort at St. Augustine as being too large, a useless parade even though adding to the beauty of the place. "The money," he observes, "would have been better laid out on roads and fences." Some of his statements dealing with the town of New Smyrna have caused bitter denials and prolonged controversy even to the present day.

In West Florida life was rougher and conveniences for living fewer than in East Florida and her neighbors to the north. Romans was so struck by the frontier simplicity which he found that he wrote: "The manners and way of life of the white people differ greatly from that in other provinces, particular in respect to clothing." He follows then with a discussion of the kinds of clothing worn by English men and women and by the natives.

Romans notes carefully the habits and character of the colonists. He comments in some detail on the prevalence of drinking. The usual drink was water tempered with a moderate quantity of West Indian rum. Among the higher classes Portuguese and Spanish wines were consumed, while New England rum, "that bane of health and happiness," was drunk by the lower classes. In discussing relations between the British manufacturer and the colony and between the colonial trader and the Indian, he observes more than once that the white men were "always more prone to savage barbarity than the savages themselves."

Like so many writers on Florida, Romans often calls attention to that "happy climate where all the seasons of the year the inestimable gifts of *Flora* and *Pomona* are common, where snow or ice are (sic) seldom seen, and where the cruel necessity of roasting one's self before a fire is utterly unknown." He describes East Florida at each season of the year, stressing the heat of summer but lauding "one of the finest climates in the world from the end of September to June. In St. Augustine people enjoy good health and live to a great age." Winds as they affect the peninsula are described in some detail. Even at this early date violent storms beginning in September are pictured as dreaded in Florida.

The style of the *History* follows no law. Frequently digressive and at times bombastic, it is nevertheless often picturesque and original. One must bear in mind that Romans was a foreigner struggling with the intricacies of the English language. He says himself of his writings, "It cannot be expected that I would excel in elegance of composition or correctness of language." Yet his facts are authentic, valuable, and interesting. He has given us a fascinating, true account of early Florida.

Since its rarity has kept many students of history and most laymen from reading it and since the knowledge of the past is a prerequisite to the understanding of the present, the publishers of this reprint are to be commended for making available this important Florida document.

Louise Richardson, A.B., M.A.

PARTIAL LIST OF CONTENTS

	Page
Aborigines	25
Acadians, Industry	79-80, 126
Agriculture	122
Arkansas Indians	68, 69
Cattle	119, 125
Chef Menteur	4
Chicasaw Indians	39-41, 44, 67
Choctaw Indians	38, 41, 47, 48, 50-56-, 208, 209
Climate	2, 5, 9, 151
Coffee Culture	113-114
Commerce, Georgia in 1754	74, 75
Corn, Indian Cultivation of	83
Cotton	97, 98
Creek Indians	49, 56, 64, 62-68, 209
Diseases	153-172
Emigration, Cost of	129-137
Fort Rosalie	222
Games, Indian	52-55
Hieroglyphics, Indian	70, 71
Hooma Chitta	222
Health	151, 154-157
Indians—	
General Description of	28-30
Funeral Customs	60, 61, 68
Habitations	45
Morality of	58
Perversion	56
Polygamy	67
Indigo, Culture of	93-96

	Page
Madder, Culture of	110-112
Longevity	8
Marshes	21, 152
Mississippi	85-108, 129-130, 234, 272
Mobile	7, 49, 81, 287-289
Mulberry Tree	99
Natchez	222
Naval Stores	143
Navigators, Instructions for	231
Negroes	77-79
New Orleans	4, 104, 147, 151
Prices of Food	80
New Smyrna	179-182
Pearl River	152
Pensacola	147, 151, 153, 175, 176, 201, 202, 278-285, 286
St. Augustine	6, 25, 174
St. John's River	25, 173, 174, 182, 183
St. Mary River	172
Rice, Culture	88-89
Silk, Culture	100
Slavery	71-73, 76-77
Tampa	24, 281-283
Tea, Culture	115-118
Timber	123, 124, 142
Tobacco, Culture	102, 103
Trees	12-15, 17-19
Witchcraft	50
Yazoo River	123
Yellow Fever	156-159

Reason without experience can do nothing; being no more than the mere dreams, phantasms, and meteors of ingenious men, who abuse their time.

There is need of much diligence and labour, before man can be thoroughly instructed.—LINNEUS.

All things contained in the compass of the universe declare, as it were with one accord, the infinite wisdom of the Creator; for whatever strikes our senses, whatever is the object of our thoughts, is contrived, as to assist in manifesting the divine glory (i.e.) the ultimate end which God proposed in all his works. Whoever duly turns his attention to the things on this our terraqueous globe, must necessarily confess, that they are so connected, so linked together, that they all tend to the same end, and to this end a vast number of intermediate ends are necessary.

—ISSAC BIBERG.

Man, the servant and explainer of nature, observes and practises as much as he has learned, concerning her order, effect, and power; further he neither knows nor can do.—BACON.

TO

JOHN ELLIS Esqr

Fellow of the Royal Societys of

LONDON and UPSAL

Agent for the Province of Weſt-Florida

This Work is with the greateſt Reſp-
= ect most humbly Dedicated : by

His most Obedient
humble Servant

Bernard Romans.

INTRODUCTION

PREFACES, at this present day, become such impertinent things, that it is almost improper to offer one without an apology.

The many different reports, which have prevailed in America, since the cession of the Floridas, concerning their state, situation and soil, joined to the natural desire of those concerned, to see a good account of those so celebrated countries, I hope will be apology enough in the present case.

Conscious of being, from experience, sufficiently enabled to give a just account of them, I have undertaken the following sketch, or out-lines of a future natural history of those countries, in hopes that some abler hand may be thereby induced to take up the pen, and furnish the world with a complete work of that kind for these provinces; being well assured, that no part of British America will furnish the naturalist with more variety.

I offer this humble attempt without any recommendations, or praises, of my own; only I beg to assure my reader, that I have, through the whole, adhered so strictly to truth, as to make no one deviation therefrom willingly, or knowingly; guarding on the one hand against the misrepresentations, wherewith the authors of the numerous and noted puffs, concerning these provinces, have so plentifully interlarded their labours; and on the other, against the prejudices of those, who have taken so much pains to render this country undeservedly despised.

No elegance of style, nor flowers of rhetoric, must be expected from a person, who is conscious that he is not sufficiently acquainted with the language, to write in such a manner as will please a critical reader, and if he has wrote so as to be intelligible, he hopes the candid will excuse such inaccuracies in composition as it is difficult for a foreigner to avoid.

LIST OF SUBSCRIBERS TO THIS WORK

A

Mr. Benjamin Andrews, Boston,
Capt. Samuel Andrews, Newbury-Port,
John Antill, Esq; New-York
Capt. Vincent P. Ashfield, ditto,
Mr. Thomas Aylwin, Boston,
Mr. Thomas Allen, New-London.

B

Mr. Theophilact Bache, New-York,
Mr. Isaac Beers, New-Haven
Charles Bernard, Esq; East Florida, 3 copies,
Capt. Robert Bethell, Philadelphia,
George Bethune, Esq; Boston,
Mr. Clement Biddle, Philadelphia,
Mr. Owen Biddle, ditto,
Captain John Blake, Boston,
Mr. William Bradford, Philadelphia,
Mr. Anthony L. Bleeker, New-York,
Honourable James Bowdoin, Esq; Boston,
Dirk Brinkerhoff, Esq; New-York,
Lieutenant Brudenell, of the Navy,
Capt. Ashbel Burnham, Middletown,
Thaddeus Burr, Esq; Fairfield,
Adam Babcock, Esq; New-Haven,

C

Captain Richard Cary, Boston,
Chamber of Commerce, 12 copies,
His Excellency Peter Chester, Esq; Governor of West-Florida,
Matthew Clarkson, Esq; for the Library Company, Philadelphia,
Captain Benjamin Cobb, Boston,
Captain Tristram Coffin, Newbury-Port,
Honourable Commissioners of his Majesty's Customs, Boston,
Capt. W. Coombs, Newbury-Port,
Capt. James Creighton, jun. New-York,
Captain William Curtis, New-York,
Messieurs Cox and Berry, Boston.

D

Capt. Benjamin Davis, New-York,
Mr. James Davis, ditto,
Lieut. Dawson, of the Navy,
Capt. Patrick Dennis, New-York,
Mr. Geradus Duyckink, ditto,
Mr. Timothy Dwight, jun. Yale College.

E

John Ellis, Esq; F. R. S. London,
George Erving, Esq; Boston,
John Ewing, D. D. Philadelphia.

F

Edmund Fanning, Esq; Secretary to Governor Tryon,
Capt. Nicholas Fletcher, New-York,
John Fothergill, M.D. F.R.S. London,
Mr. Philip Francis, Philadelphia,
Mr. David Frazier, Sussex, New Jersey,
Capt. Benjamin French, Lansingburgh,
Capt. Joseph French, New-York.

G

Dr. Sylvester Gardiner, Boston,
Capt. Martin Gay, ditto,
John Gibson, Esq; Philadelphia,
Capt. John Gore, Boston,
Joseph Green, Esq; Boston,
John Griffith, Esq; New-York,
Mr. Anthony Griffiths, do.
The right Noble Gronovius, one of the Deputies from the ancient city of Leyden, to the Chambers of Finances at the Hague.

H

Mr. Reuben Haines, Philadelphia,
His Excellency Major General Frederick Haldiman, Esq;
Mr. Willis Hall, Boston,
Jonathan Hampton, Esq; Elizabeth-Town,
Honourable John Hancock, Esq; Boston,
Ditto for Harvard College,
Capt. J. Harrison, of the Ship Queen, S. Caro.
Capt. William Henderson, Middletown,
Samuel Holland, Esq; Surveyor General for the Northern District,
Stephen Hooper, Esq; Newbury-Port,
Thomas Howell, Esq; New-Haven,
Capt. Francis Hutcheson, 60th Regiment,
Lieutenant Hunter, of the Navy,
Capt. John Hylton, New-York.

I

Capt. Jabez Johnson, New-York,
Evan Jones, Esq; Pensacola,
Mr. Ralph Isaacs, New-Haven.

K

Mr. Henry Knox, Boston

L

James de Lancey, Esq; New-York,
Mr. John Landon, Boston,
Mr. William Lewis, Esq; New-York,
Leonard Lispenard, Esq; do.
Mr. Abraham Livingston, jun, do.
Hon. Philip Livingston, Esq; Pensacola,
Mr. James Lockwod, New-Haven,
John Lorimer Esq; M.D. Pensacola,
Mr. Samuel Loudon, New-York,

John Lukens, Esq; Philadelphia, 6 copies,
Mr. Char. Lukens, York-Town, Pennsylvania,
Thomas Lynch, Esq; Charles-Town, S. Caro.
Christopher Leffingwell, Esq; Norwich.

M

Marine Society, New-York,
Marine Society, of Boston,
Marine Society, of Salem,
Marine Society, of Newbury-port,
Mr. Edward M'Michael, Sussex, New-Jersey,
Capt. Alexander M'Dougall, New-York,
Mr. William Malcom, ditto,
Richard Martin, Esq; Rio Bueno Jamaica,
Thomas Martin, Esq; Portsmouth, N. Hampsh.
Mr. Charles Marshal, Philadelphia,
Mr. Samuel Miles, ditto,
Capt. Magnus Miller, ditto,
Mr. John Minshall, New-York,
Hon. J. Montague, Esq; Rear Adm. of the Blue,
Capt. Montresor, as Successor to Capt. T. Sowers, six Copies for the Engineers Office, America,
Capt. Thomas Moore, New-York,
Mr. Ph. Moore, Philadelphia,
Capt. Roderick Morrison, Newbury-port,
Robert Morris, Esq; Philadelphia.

N

J. M. Nesbitt, Esq; Philadelphia,
Capt. Samuel Newhall, Newbury-Port,
Capt. Downham Newtown, N. Providence,
Capt. James Nicoll, Newbury-Port,
Capt. Silas Nowell, ditto.

P

Major Adino Paddock, Boston,
Mr. Z. Parsons, Springfield,
M. Timothy Penny, Boston,
Mr. Isaac Green Pearson, Newbury-Port,
Mr. Joseph Pemberton, Philadelphia,
William Philips, Esq; Boston,
Mr. Samuel Philips, ditto,
John Pitts, Esq; ditto,
Mr. Peter le Pool, Charlestown, South-Carolina, 6 Copies,
William Powell, Esq; Boston,
Capt. Job Prince, ditto,
Mr. John Perrit, Norwich.

R

Capt. Thomas Randall, New-York,
Mr. Gerrit Rapalje, do.
John Rapalje, Esq; do.
Messieurs Read and Yates, do.
Capt. John Rionson, Boston,
Mr. Huybertus Romans, Amsterdam,
John L. C. Roome, Esq; New-York,
Parr Ross, Esq; N. Providence,
John Rowe, Esq; Boston,
Thomas Russel, Esq; Charlestown, New-England,

S

Capt. Giles Sage, Middletown,
Daniel Sargeant, Esq; Cape Ann,
Mr. Elias Shipman, New-Haven,
Joseph Shebourne, Esq; Boston,
Jonathan Simpson, Esq; ditto.
Major John Small, ditto,
Mr. Christopher Smith, New-York,
Mr. Archibald Stewart, Sussex, New-Jersey,
Mr. H. W. Stiegel, Lancaster County, Pennsylvania, 6 Copies,
Ebenezer Storer, Esq; Boston,
Capt. Symonds, of the Navy,
Mr. Nathaniel Shaw, jun. New-London.

T

Mr. Nathaniel Tracey, Newbury-Port, 2 Copies,
Mr. John Tracy, Newbury-Port,
Mr. Robert Tracey, ditto,
His Excellency William Tryon, Esq; Governor of N. York,
William Todd, Esq; York, in Old England,
Capt. Samuel Tuder, New-York.

V

Mr. Anthony Van Dam, for the New-York Insurance Office,
Mr. John Van Rensselaer, jun. Albany,
William Vassal, Esq; Boston.
John Vassal, Esq; ditto,

W

Capt. Jeremiah Wandsworth, Middletown,
Hon. Hugh Wallace, Esq; New-York,
Mr. Joshua Wallace, Philadelphia,
Oliver Wendell, Esq; Boston,
Mr. Joseph Webb, Weatherfield,
Mr. Samuel Webb, ditto,
Edmund Rush Wegg, Esq; Pensacola,
Joseph Wharton, Esq; jun. Philadelphia, 3 Copies,
Hon. Thomas Willing, Esq; ditto,
Capt. Erasmus Williams, New-York,
Mr. Jonathan Williams, tertius, Boston,
Mr. Thomas C. Williams, Philadelphia,
Mr. William Wilson, New-Providence,
Capt. Edward Wigglesworth, Newbury-Port,
Capt. Aaron Willard, Boston,
Capt. Is. L. Winn, New-York,
Mr. Joshua Winslow, Boston,
Mr. Joseph Whitall, Philadelphia,
Capt. James Wright, New-York,
Capt. William Wyer, Newbury-Port.

Y

Robert William Yates, Esq; Albany,
Dr. Thomas Young, New-Port, Rhode-Island.

SUBSCRIBERS FOR THE BOOK ONLY

Mr. John Adams, Boston,
Mr. Joseph Barnell, ditto,
Mr. Samuel Biagdea, New-Haven,
James Burrows, Esq; Boston,
Rev. Mr. Carey, ditto,
Messieurs Cox and Berry, ditto, 25 Copies,
Mr. Thomas Fanning, ditto,
Mr. D. S. Franks, Quebec,
Dr. John Greenleaf, Boston,
Mr. Roger Haldane, New-Haven,
Mr. William Kennedy, Boston,
Mr. Henry Knox, ditto, 50 Copies,
Mr. James Lockwood, New-Haven, 25 Copies,
Mr. William Molineaux, Boston,
Mr. Henry Pelham, ditto,
Capt. Rufus Putnam, Brookfield, 6 Copies,
Dr. Isaac Rand, Boston,
Hector St. John, Esq; Grey Court,
Mr. William Tuder, jun. Boston.

A Concise Natural History of East and West-Florida

A describer of countries, ought in a great measure, to imitate a building Engineer, in first laying before those, whom he will employ, accurate and distinct plans of his intended work, thereby enabling them to judge more distinctly of the execution thereof. I think that in a work of this nature, I could not do this better than by directing my readers to the charts or plans accompanying it, in which they will undoubtedly find materials to form just ideas of the places herein described.

To reduce my work to some regularity, I shall proceed from the East, Westward, and begin with the Peninsula, dividing it into two parts, which I will call climates, the one beginning at Amelia or St. Mary's inlet, in latitude 31: and extending southward to the latitude of 27: 40: this will include the rivers St. Mary, Nassau, St. John's or Ylacco, and the Musketo Lagoon (for surely no one can call this last a river) besides several smaller ones, which will be mentioned in their places; these all empty themselves on the Eastern side of the Apalachicola (the boundary between the two Floridas) the Oskaulaskna, the Apalachian, St. Juan de Guacaro, vuglarly called little Seguana, the river Amaxura, and the Manatee, which last falls into the bay of Tampe, or harbour of Spirito Santo, and which I have first discovered.

The other, or Southern climate, beginning at the latitude 27: 40: and extending Southward to the latitude of 25, on the main, or to 24: 17: including the keys; this contains a large river, which empties itself into the new harbour, of which I am the first explorer, we have given it the name of Charlotte harbour, but neither harbour nor river have been described by the Spaniards in their maps, and the Spanish fishermen distinguish the place by the names of its inlets, which are five, and will hereafter be described; next is Carlos bay and Carlos harbour, into which the river Coloosahatcha empties itself; further South are not any more deserving the name of rivers, but such as they are, I shall give them a place also; on the East side is only the river

St. Lucia, with its Southern branch, the river Ratones, and the Lagoon, known by the name of Aisa Hatcha, Rio d'ais, or Indian river, some others can scarcely be ranked among rivers, but will likewise be more particularly mentioned hereafter.

After this general division of the country, I think it is not improper to begin with an account of the air, which this province enjoys very pure and clear fogs are seldom known any where except upon St. John's river, but the dews are very heavy, the spring and summer are in general dry, the autumn very changeable; the beginning of winter wet and stormy, but the latter part very dry and serene; from the end of September to the end of June, there is perhaps not any where a more delightful climate to be found, but all July, August, and most of September are excessively hot, yet the changes from hot to cold are not so sudden, as in Carolina, and frost is not frequently known, the noon day's sun is always warm, the severest cold ever known there affects not the tender china orange trees, which grow here to a very great perfection, I scruple not to say, that this fruit here exceeds in goodness every other of the kind I have yet seen, however the change from the middle of this climate, to the Northern part of it is much more perceptible from heat to cold, than it is to the southward from cold to heat, in the year 1770 and 1771, I felt. Very severe weather about the river Nassau. To the southward of the town of St. Augustine, the climate changes so gradually, that it is not perceivable to the above named lat. of 27: 40: where there is no frost at all, and which I have always set down as the line of no frost. From this line to the southern extent is a most charming climate, the air almost always serene; on the east side the common trade wind, and on the west side the Apalachian sea breeze from the west to the north-west, refresh this delightful Peninsula during the summer; here we find all the produce of more northern climes mixed with the inhabitants of the Tropics, and this as well in the water as on the land, nor is there ever so great a cold as to destroy the fruits of the south, nor so great a heat as to parch the produce of the north; in all this Peninsula it is remarkable, that rain is always prognosticated one or two days before it falls, and this by either an immoderate dew or no dew at all, so that if a very heavy dew falls, it is a certain sign of rain, and the same if on a calm fine night, there be no dew, but I cannot account for this phenomenon.

The winds are not so very changeable here as they are further to the northward, but are during the greatest part of spring, the whole summer, and beginning of autumn, generally between the east and south east, and during the last of autumn, and first part of winter, they are commonly in the north east quarter; the latter part of the winter, and first of spring they are more generally west and north west, the autumnal equinox is to be dreaded here, as well two or three weeks before, as two or three months after it, great storms will then happen, and many vessels are drove on shore, or otherwise disabled: I have never heard of much mischief in the vernal equinox, and if a hurricane was ever known in this Peninsula, it was on the 29th of October 1769, when there was a terrible gust between the lat. 25: 10, and 25: 50, which blew many trees down, and drove the Snow Ledbury a shore, where she remained dry on a key, now distinguished by her name, but heretofore considered as a part of what was improperly called by the name of Key Largo.

The fatal hurricane of August 30, 31, September 1, 2, 3, anno 1772, was severely felt in West Florida, it destroyed the woods for about 30 miles from the sea coast in a terrible manner, what were its effects in the unsettled countries to the eastward, we cannot learn; in Pensacola it did little or no mischief except the breaking down of all the wharfs but one; but farther westward, it was terrible; at Mobile every thing was in confusion, vessels, boats, and loggs were drove up into the streets a great distance, the gullies and hollows as well as all the lower grounds of this town were so filled with loggs, that many of the inhabitants got the greatest part of their yearly provision of firewood there; all the vegetables were burned up by the salt water, which was by the violence of the wind, carried over the town, so as at the distance of half a mile, it was seen to fall like rain; all the lower floors of the houses were covered with water, but no houses were hurt except one, which stood at the water side, in which lived a joiner, a schooner drove upon it, and they alternately destroyed each other; but the greatest fury of it was spent on the neighbourhood of the Pasca Oocolo river; the plantation of Mr. Krebs there was almost totally destroyed, of a fine crop of rice, and a large one of corn were scarcely left any remains, the houses were left uncovered, his smith's shop was almost all washed away, all his works and out houses blown down; and for thirty miles up a branch of this river which (on account of the abundance of that species of cypress* vul-

---
*Cupressus Thyoides.

garly called white cedar) is called cedar river, there was scarce a tree left standing, the pines were blown down or broke, and those which had not intirely yielded to this violence, were so twisted, that they might be compared to ropes; at Botereaux's cow pen, the people were above six weeks consulting on a method of finding and bringing home their cattle; twelve miles up the river, live some Germans who, seeing the water rise with so incredible a rapidity, were almost embarked, fearing an universal flood, but the water not rising over their land, they did not proceed on their intended journey to the Chactaw nation. At Yoani, in this nation, I am told the effects were perceivable; in all this tract of coast and country the wind had ranged between the south south east and east, but farther west its fury was between the north north east and east, a schooner belonging to the government having a detachment of the sixteenth regiment on board, was drove by accident to the westward as far as Cat Island, where she lay at an anchor under the west point, the water rose so high, that when she parted her cables, she floated over the island, the wind north by east, or thereabout she was forced upon the Free masons islands, and lay about 6 weeks before she was got off, and if they had not accidentally been discovered by a hunting boat, the people might have remained there and died for want, particularly as water failed them already when discovered; the effect of this different direction of the current of air or wind was here surprising, the south easterly wind having drove the water in immense quantities up all the rivers, bays, and sounds to the westward, being here counteracted by the northerly wind, this body of water was violently forced into the bay of Spirito Santo at the back of the Chandeleurs, Grand Gozier, and Breton Isles, and not finding sufficient vent up the rigolets, nor down the outlets of the bay, it forced a number of very deep channels through these islands, cutting them into a great number of small islands. The high island of the Chandeleur had all the surface of its ground washed off, and I really think, had not the clay been held fast by the roots of the black mangrove, and in some places the myrtle (Myrica) there would have been scarce a vestige of the island left; at the mouth of Mississippi all the shipping was drove into the marshes; a Spanish brig foundered and parted, and a large crew was lost, some of the people were taken from a piece of her at sea, by a sloop from Pensacola a few days after; in the lakes at Chef Menteur, and in the passes of the rigolets, the water rose prodigiously and covered the low islands there two feet; at St. John's Creek, and New Orleans, the tide was thought extraordinary high, but at all these last places there was no wind felt, being a fine serene day with a small air from the eastward.

The most extraordinary effect of this hurricane was the production of a second crop of leaves and fruit on all the mulberry trees in this country, a circumstance into which I very carefully enquired, but could not learn from the oldest and most curious observers that this had ever happened before; this tardy tree budded, foliated blossomed. and bore ripe fruit with the amazing rapidity of only four weeks time immediately after the gust, and no other trees were thus affected.

The south and south west winds make a thick heavy air, and are in my opinion hurtful to the lungs; they also occasion the sultry weather, so much complained of in July and August. The winds from the eastern quarter every where between the south east and the north east, are cool and moist, and they cause the frequent showers, by which the very sand of this climate is endued with so prodigious a vegetative power that it amazes every one. The winds from the east to the north are agreeably cool, and from the north to the north west, occasion what is here called cold weather; I have frequently kept thermometrical journals, but have none left now for inspection.

I remember the general height of the mercury on Fahrenheit's scale, to have been, in the shade where the air was not prevented circulating freely about it, between 84° and 88° and on some sultry hot days in July and August, I have known it to rise up to 94°, when at the same time by carrying it out and exposing it to the sun, it will rise in a very short time up to 114°, nor can I remember ever to have seen it above one or two degrees below the freezing point; it is impossible for one to imagine how inexpressibly temperate the weather is here from the latter end of September to the latter end of June; the western part of this northern division is not so very hot in summer, as the whole eastern shore of the Peninsula is, but its sea shore is much more exposed to the bleak winter winds.

In the southern division I have never seen the mercury in Fahrenheit's thermometer below the temperate point, and I cannot remember ever to have seen it higher than in the northern division.

This southern part of the Peninsula is in the months of May, June, July, and August very subject, on its west side, to dreadful squalls, and there is a certainty of one or more of these tornadoes every day, when during that season, the wind comes any where between the south south east, and south west, but they are of very short duration; then also thunder and lightning is frequent, but nothing near so violent as in Carolina and Georgia, nor do I remember any more than one instance of damage occasioned by it, when it made a large hole in a stone wall of a house at St. Augustine; yet very few electrical conductors are made use of there.

Before I quit this subject of the air, I cannot help taking notice of a remark, which I have read some where, made by Dr. James McKenzie, which is that dampness or discoloring of plaister, and wainscoat, the soon moulding of bread, moistness of spunge, dissolution of loaf sugar, rusting of metals, and rotting of furniture, are certain marks of a bad air; now every one of those marks except the last, are more to be seen at St. Augustine, than in any place I ever was at, and yet I do not think, that on all the continent, there is a more healthy spot; burials have been less frequent here, than any where else, where an equal number of inhabitants is to be found, and it was remarked during my stay there, that when a detachment of the royal regiment of artillery once arrived there in a sickly state, none of the inhabitants caught the contagion, and the troops themselves soon recruited; I also know of several asthmatic and consumptive subjects, who have been greatly relieved there; the Spanish inhabitants lived here to a great age, and certain it is, that the people of the Havannah looked on it as their Montpelier, frequenting it for the sake of health; I therefore ascribe the above circumstances to the nature of the stone, wherewith the houses are built.

Haloes, or as they are vulgarly called circles round the sun and moon, are very often seen, and are sure forerunners of rain if not wind storms; those of the sun are less frequent, but they are always followed by very violent gales of winds; it is remarkable, that if in those haloes a break is observed, that break is always towards the quarter, from whence the wind begins; water spouts are often seen along this coast, but I cannot learn that they ever occasioned any mischief, nor could I learn, that earthquakes have been experienced in this part of the world.

Of West Florida, there needs scarce any thing more to be said, with regard to the article of climate, or air, than what I have said of my northern division of East Florida, it agreeing in every respect therewith, except that the winter is something more severe, it often killing tender fruit trees;* however, as the sickness of 1765 at Mo-

---
*In 1771-2, it killed apple and pear trees.

bile, has been a subject of much discourse, and as it has been set up (by people who would if possible prevent the population of so fine a country) as a scarecrow to such, as are easily deceived by appearances, and never enquire deeper than external shews; this fatal disorder has been followed by the entire ruin of Mobile, and had nearly spoiled the reputation of Pensacola, which though situate in as fine, airy, dry and healthy a site as any on the continent, and at least at a distance of sixty miles from Mobile, had yet the misfortune to be confounded with it, and to be thought liable to the same misfortunes; I will give as faithful an account of that illness, as has come within the verge of my knowledge.

Mobile was originally built by the French, after they had left their old Fort Condè, thirty miles higher up the Tombecbé, having found that situation very inconvenient; they now made at least as injudicious a choice in another respect, by placing themselves at a distance from good water, on low ground, and directly opposite to some marshy islands, at the division between the salt and fresh water, a situation well known in America not to be eligible for the sake of health, but the convenience of the navigation up to it being the best in their possession at that time, its being a barrier against the Spaniards, and the easy communication with the Chactaw and Upper Creek nations, as well as with the Mississippi, made people forget the evils attending it, and it soon became, from a fort, a pretty town, with some very good houses built in no inelegant taste, yet the French inhabitants duly observing the inconveniences of this unhealthy spot, adapted their constitutions to it, by a regular sober life, being uncommonly careful to get their drinking water from a rivulet at the distance of three miles, where it is very good, neither did they give into excess of drinking spirituous liquors and wine, and at the season, when the continued heat caused a putrefaction of the water in pools, and exhaled the moisture of this low ground, thereby filling the air with noxious vapours, and thus occasioning the acute epidemical disorders (that proved so fatal in the year 1765) those prudent inhabitants retired to their plantations up or down the river, some even at a small distance, there to enjoy a freer circulation of a less putrified air, thus also by the depopulation of the town, the remaining inhabitants suffered less by being less crowded together, and there was such instances of longevity here as are not to be outdone in any part of America. Let me beg leave to mention among many others, one more commonly known, it is the Chevalier de Lucere's family, who are now all very old, and whose mother not many years since died by breaking one of her legs, that had been so much calcarizated by the gout, that it snapped by stepping into bed, she died aged far above one hundred years. One other I shall mention, more familiar to me, which is that of one Mr. François, who lives now about five miles below the river Poule: In September 1771, I called there, the old man told me he was then past eighty three years of age, that the old woman, whom I saw putting bread into the oven, was his mother; and that she was one of the first women that came from France to this country; I saw her about her domestick business in many ways; in a very cheerful manner, singing and running from place to place as briskly as a girl of twenty; Mr. François told me, that at the age of sixty he fell out of a pine tree, above fifty feet high, with his loins over a fallen one, that he with difficulty recovered, and that had it not been for that accident, he would not, as he thinks, yet have been sensible of the heavy hand of time; that he was still a hearty cheerful old man, was evidently to be seen; when I came to the river Poule in October 1772, I met the same old gentleman fishing at the mouth of the river, on my asking him whether this diversion was agreeable to him, he told me, that his mother had an inclination to eat fish, and he was come to get her a mess; he was then on foot and had five miles to come to this place, and as much back with his prey, after catching it; a very dutiful son this at eighty five! He lives comfortably at an agreeable place, and on the produce of a midling large stock of cattle.

Many more of this kind might be mentioned, but these two being more universally known, I chose to relate them only. Far otherwise was it with our sons of incontinence, who upon their arrival, and after their first taking possession of this country, lived there so fast, that their race was too soon scampered over; midnight carouzals, and the converting day into night, and night into day was all the study of those gay, those thoughtless men, who sported with their lives, as with a toy not worth esteeming; the fatal effects of their debauches; joined to the consequences of the situation of their residence, made their lives indeed comparable to grass, flourishing to day, and withering to morrow; but as if a punishment for this abandoned life, was not sufficiently incurred by its own fatality, in the year 1765 arrived a regiment (I think the twenty first) from Jamaica, with them they brought a contagious distemper; contracted either in the island, or on their passage; these men, like most soldiers; lived a life of intemperance, and besides, drank the water out of the stagnated pools, which I myself have even in the winter, seen such as to fill a man with horror at the thought of making use thereof, this and other inconveniences of a soldier's life, joined to their arriving in a bad season, swept them off so as scarce to leave a living one to bury the dead. See there the true reason of the sickly character of the climate, and of the destruction of this once flourishing town, whose situation by far exceeds that of Savannah in Georgia, in every respect.

It is an almost invariable rule for people, who intend going to a different climate, to consult some friend or acquaintance on the manner of life, he would advise him to lead, I have never yet heard of one going to Florida, who was not told by his friend, that a free glass was necessary; how true this is, I shall not pretend to say, but certain it is, that the advice is almost always too freely followed, the free glass generally degenerating into a glass of excess.

Notwithstanding all I have above asserted, it is not to be denied, that during the hot months, the air is not so wholesome as in the other season, but even then it does not so much affect careful strangers, and new comers, as those who have been some time there and live irregular lives.

The night air is not so much to be dreaded here, as in countries where the sun is vertical, or nearly so, and consequently, by its long absence, makes a chilling penetrating night follow a burning day, but here it is not long enough absent to cool the atmosphere sufficiently to hurt the unwary sleeper, who during the first heat of a sultry night perhaps has exposed his open pores to the mercy of the air.

The atmosphere is, during this season, so burning hot, that undoubtedly very sudden rarefactions of the humours are often experienced, which cause such abundant perspiration, that water, as soon as drank, pervades the open pores, so that the human skin seems to be comparable to a wet spunge when squeezed; yet although the water is here very cool (and if it has not this quality naturally, it is artificially made to acquire it) we never hear of the fatal effects of water drinking, so often experienced in the cities of New York and

Philadelphia, the reason perhaps is, that it is seldom if ever drank unmixed.

I will however venture to foretell, that on opening the woods of this country for cultivation, which will naturally drain ponds, gullies, &c. the air will be here very little affected by those pernicious vapours, which have so uncommon an influence over the humours and fibrous parts of the human frame, as to destroy their harmonious concordance (may I be admitted the phrase?) and occasioning them to relax, and thereby producing weaknesses, lassitudes, and finally dangerous and fatal disorders.

If we consider the effects of heat and humidity on the hardest substances, such as wood, and even metals, which are thereby expanded, and have the union of their solid parts relaxed, it may give us an idea, how much more their effects must be felt in the animal oeconomy at times, when fire and water unite their dissolving powers to act on all nature.

A very dry hot air, though less dangerous to the body, than a hot moist one, has yet very nearly the same effects, as it partially dries the Ponds, Marshes, Swamps, &c. leaving the remaining water and effluvia of small animals to exhale, and spread their noxious vapours through the atmosphere.

Every inhabitant of any part of America knows, that the sudden transitions from cold to heat so prevalent on that continent, are much more to be dreaded, than any of the above named causes of immoderate heat, cold, moisture, and drought.

I am now to consider the nature and appearance of the earth, which in this part of America, may be divided into six different sorts, much the same as in Carolina, with this distinction, that it is much more unequally divided.

I shall treat of them by the names of pine land, Hammock land, savannahs, swamps, marshes, and bay, or cypress galls.

First the pine land, commonly called pine barren, which makes up the largest body by far, the Peninsula being scarce any thing else; but about an hundred miles towards the north west from St. Augustine, and about two hundred from the sea in West Florida, carry us intirely out of it. This land consists of a grey, or white sand, and in many places of a red or yellow gravel; it produces a great variety of shrubs or plants, of which I shall hereafter describe some, the principal produce from whence it derives its name in the *pinus foliis longissimis ex una theca ternis,* or yellow pine and pitch pine tree, which I take to be a variety of the same species, both excellent and good timber.

Also the *chamaerops frondibus palmatis plicatis stipitibus ferratis,* of whose fruit all animals are very fond.

It is on this kind of land, that immense flocks of cattle are maintained, although the most natural grass on this soil is of a very harsh nature, and the cattle not at all fond of it, it is known by the name of wire grass; and they only eat it while young; for the procuring it young or renewing this kind of pasture, the woods are frequently fired, and at different seasons, in order to have a succession of young grass, but the savannahs that are interspersed in this kind of land furnish a more plentiful and more proper food for the cattle.

Some high pine hills are so covered with two or three varieties of the *quercus* or oak as to make an underwood to the lofty pines; and a species of dwarf chesnut is often found here; another species of a larger growth is also found in the lower parts, particularly in the edges of the bay or cypress galls.

This barren and unfavourable soil in a wet season bears many things far beyond expectation; and is very useful for the cultivation of peach and mulberry orchards; this land might also be rendered useful for many other purposes, but either the people do not choose to go out of the old beaten track, or content themselves with looking elsewhere for new land improveable with less cost; the method of meliorating it is certainly obvious to the meanest capacity, as it every where, at a greater or less depth, covers a stiff marly kind of clay, which I am certain, was it properly mixed with the land, would render it fertile, and this might be done with little expence, the clay laying in some places within half a foot or a foot of the surface; in most places it is found at the depth of three, four, or five feet, consequently not very hard to come at. In East Florida, in the southern parts, this kind of land is often very rocky, but especially from the latitude 25: 50, southward to the point, where it is a solid rock, of a kind of lime stone covered with innumerable small, loose and sharp stones, every where.

In West Florida the pine land is also frequently found rocky, with an iron stone, especially near where the pines are found growing in a gravelly tract, which is frequently the case here.

The hammock land so called from its appearing in tufts among the lofty pines; some small spots of this kind, if seen at a distance, have a very romantick appearance; the large parcels of it often divide swamps, creeks, or rivers from the pine land, this is indeed its most common situation; the whole of the up lands, remote from the sea in the northern parts, is this kind of land, its soil is various, in some places a sand of divers colours, and in East Florida, often a white sand; but the true hammock soil is a mixture of clay and a blackish sand, and in some spots a kind of ochre, in East Florida some of this is also sometimes found rocky; on every kind of this land lays a stratum of black mould, made by the decayed leaves &c. of the wood and other plants growing upon it; the salts contained in this stratum render it very fruitful, and when cleared this is the best; nay the only fit land for the production of indigo, potatoes, and pulse; the first crops, by means of the manure above mentioned, generally are very plentiful, but the salts being soon evaporated, if the soil over which it lay, should prove to be sand, it is not better than pine land; the other sort bears many years planting; its natural produce is so various in this climate, that the compleat description of all, would be more work than one man's life time would be sufficient for, the principal however are the following:

Quercus alba Virginiana.	Virginian white oak.
Quercus alba pumilis.	Dwarf white oak, or post oak.
Quercus, foliis oblongis non sinuatis, semper virens.	Evergreen oak with oblong entire leaves, or live oak.
Quercus nigra, folio non serrato, in summitate quasi triangulo.	Black oak, with leaves serrated, and their tops almost triangular.
Quercus nigra foliis cunei forma, obsolete trilobis.	Black oak, with wedge shaped leaves, and having imperfectly three lobes.
Quercus nigra Marilandica, folio trifido, ad sassafras accidente.	Black Maryland oak, with trifid leaves resembling sassafras.
Quercus rubra Carolinensis, verens muricata.	Carolina red oak, prickly when young.
Quercus castaneae foliis, procera arbor.	Chesnut leaved oak of a large size.

Juglans alba, fructu ovato compresso profunde insculpto durissimo, cavitate intus minima.

White walnut, or hickory with egg shaped f r u i t closely grasped, and buried, in a very hard shell, with the smallest inward cavities.

Juglans Virginiana alba minor.

Small Virginian hickory tree.

Nux Juglans nigra.

Black walnut.

Fagus humilis (seu castanea, pumila) racemosa fructu parvo; in capsulis echinatis, singulo.

Smallest fagus (or dwarf chesnut) having the fruit in bunches, and contained singly in a prickly pod, vulgo chinkapin.

Fagus foliis lanceolatis ovatis, Acute serratis subtus tomentosis amentis filiformi nodosis, fructu in capsulis echinatis, duplice.

Fagus, with leaves between egg and spear shaped, sharply serrated and woolly underneath, slender knotty catkins, and a double fruit in a prickly pod.

Morus foliis subtus tomentosis amentis longis, diaecis.

Mulberry with the under part of the leaves woolly, having long catkins; and trees of different sexes.

Morus, loti arboris instar, ramosa; foliis amplissimis.

Mulberry resembling the lote-tree full of branches; and large leaves.

Morus foliis palmatis, cortice filamentosa, fructu nigro, radice tinctoria.

Mulberry with hand shaped leaves a thready bark, black fruit, and the root containing a dye.

Diospyros guajacana.

Parsimmon.

Liquidambar, styraciflua; aceris folio.

Maple leaved liquidambar, yielding storax or sweet gum.

Borassus frondibus palmatis (seu) palma coccifera latifolio, fructu atro purpureo, omnium minimo.

Borassus, with hand or fan shaped leaves (or) scarlet yielding palm, with broad leaves, and a deep purple fruit which is the least of all.

Palma humilis (seu) chamaeriphis

Dwarf palm, or chamaeriphis.

Laurus foliis acuminatis, baccis caeruleis, pedicellis longis rubris infidentibus.

Laurus (seu cinnamomum sylvestre) Americana

Laurus (seu) cornus mas odorata, folio trefido, margine plano ,sassafras dicta.

Liliodendron tulipifera, tripartito, aceris folio, media lacinia velut abscissa

Magnolia maximo flore, foliis subtus ferrugineis.

Magnolia glauca laurifolio subtus albicante

Magnolia flore albo, folio majore acuminato haud albicante.

Magnolia tripetala amplissimo flore albo, fructu coccineo.

Citrus (seu) malus autantia acida.

*Illicium floridanum (seu) anisum stellatum.

Kalmia, foliis glabris lanceolatis; et corolla campanulae hypocraterique forma.

Ficus Americana, citri folio, fructu parvo purpureo.

Coccoloba (seu) prunus maritima racemosa, folio subrotundo, venoso fructu caeruleo quasi purpureo.

Laurel, with pointed leaves, and blue berries, fitting on long red foot stalks.

The wild American cinnamon Laurel

The male scented cornel or laurel tree, with a trifid leaf, having plain edges, called sassafras.

Tulip bearing liliodendron, with tripartite maple leaf, having the middle piece seemingly cut off.

Magnolia, with the largest flower, and the lower side of the leaves ferrugineous.

Magnolia, with a grey laurel leaf whitish below.

Magnolia, with a white flower, a larger pointed leaf, and not whitish.

Magnolia, with a very large white flower of three petals; and a scarlet fruit.

The sour orange.

Starry aniseed, or skimmi.

Kalmia, with smooth lanceolate leaves, and a corolla between salver and bell shaped.

American fig, with a citron leaf and a small purple fruit.

Coccoloba, or sea side plumb, growing in bunches, an almost round veined leaf, &c. the fruit blue, inclined to purple.

---

*First found growing near Pensacola, by a free Negro (Pompey) formerly belonging to Chief Justice Clifton; which Negro in his own way is a curious herbalist.

Coccoloba foliis oblongis ovatis venosis, uvis minoribus corinthiacis.

Coccoloba, with oblong egg shaped veined leaves, with pointed grape like fruit less than currants.

Zantoxylum spinosum album, quasi fraxini folio, evonymi fructu capsuleri.

Tooth a c h tree, with white spines almost an ash leaf, and the capsulum like the fruit of the spindle tree.

The savannah's are in this country of two very different kinds, the one is to be found in the pine lands; and notwithstanding the black appearance of the soil, they are as much a white sand as the higher lands round them; true it is that clay is very often much nearer to their surface, than in the higher pine lands; they are a kind of sinks or drains to those higher lands, and their low situation only prevents the growth of pines in them. In wet weather the roads leading through them are almost impassable. On account of their producing some species of grass of a better kind than the wire grass, they are very often stiled meadows, and I believe, if they could be improved by draining them, without taking away all their moisture, very useful grass might be raised in them; but on draining them completely, they prove to be as arrant a sand as any in this country. These savannahs often have spots in them more low than common, and filled with water; they are over grown with different species of the *crataegus,* or hawthorn, as also very often a species of shrub much resembling the *Laurus* in appearance, but as I never had an opportunity of seeing it in blossom, I cannot describe it, so as to ascertain the genus it belongs to; in its fruit it is widely different from any of the laurel kind, that have fallen under my inspection; it is a bacca with several cells full of an agreeable acid like the common lime from the West Indies; it is of the size of a large pigeon's egg, but more oblong; we also find it on the low banks of rivers in Georgia, and know it by the name of the Ogeechee lime. The other savannahs differ very widely from these, and are chiefly to be found in West Florida, they consist of a high ground often with small gentle risings in them, some are of a vast extent, and on the west of Mississippi, they are said to be many days journey over, the largest within my knowledge is on the road from the Chactaw to the Chicasaw nation, and is in length near forty miles over from north to south, and from one end to the other, a horizon, similar to that at sea, appears; there is generally a rivulet at one or other, or at each end of the savannahs, and some come to the river banks; in one or two of them I have seen some very small remains of ancient huts, by which I judge, they were formerly inhabited by indians; the soil here is very fertile; in some I have seen fossil shells in great numbers, in others, flint, in others again some chalk and marl; it is remarkable, that cattle are very fond of the grasses growing here; the Chicasaw oldfield, as it is termed, is a clear demonstration of this, for the cattle will come to it from any distance, even when the grass scarcely appears; and in all the circumjacent tract, are abundance of both winter and summer canes to be found on which they might more luxuriously feed. In these savannahs if a well or pond is dug, the water has a very strong nitrous taste. I have seen some very curious plants in this kind of ground, but there was no time for my examining any of them, except a nondescript of the genus *Tagetes* of a fine crimson colour. I shall in some measure describe and give the figure of this plant. The only high growth I have seen in these savannahs are some willows and other aquatic plants, by the side of rivulets, in or near them; some of the smaller kind of oaks and a few small *junipers* are also to be seen in those places, the *fragaria* or strawberry is very common in them.

Swamps are also found of two kinds, river and inland swamps, those on the rivers are justly esteemed the most valuable, and the more so, if they are in the tide way, because then the river water may be at pleasure let on or kept out, with much less labour and expence than in the other kinds; these lands are the sources of riches in these provinces, because where they lay between the sandy pine barrens, they produce that valuable staple Rice, and on the Mississippi (where much of this river land is situated a great deal higher, than the common run of it in Carolina, and other similar countries) this soil is the best adapted for corn and indigo, yet known; some of these grounds are clay, others sand, and others again partake of both; when used for rice, it matters not which of these soils they are made up of , but I believe, were the sandy ones to be quite drained, they would prove barren enough; the use of water on rice is more to suppress the growth of noxious weeds and grass, which would otherwise stifle the grain, than for promoting the growth of the rice itself, for none of the grasses can stand the water, but rice does, as long as it is not totally immersed, therefore it is, that after weeding, the planter (if he has it convenient) lets on water to about half the height of his grain; by swamps then in general is to be understood any low ground subject to inundations, distinguished from marshes, in having a large growth of timber, and much underwood, canes, reeds, wythes, vines, briars and such like, so matted together, that they are in a great measure impenetrable to man or beast; the produce of these swamps if sandy is more generally the cypress tree, which is here of three species; two of these grow in this kind of land; the common sort grows to an enormous size, but none so large, as what is seen on or near the banks of the Mississippi, the other kind vulgarly miscalled white cedar, is in great quantities near Pensacola, particularly in the swamps of Chester River; this likewise grows to a tree which may be ranked among those of the first magnitude: If these swamps are not altogether sand, but mixed with clay, and other earth, their produce is in general

Cupressus Americana foliis deciduis.	American deciduous cypress.
Cupressus semper virens seu cupressus Thyoides.	Evergreen cypress, vulgo white cedar.
Quercus alba aquatica salicis folio breviore.	White swamp oak, with a short willow leaf, vulgo water oak.
Quercus folio longo angusto salicis.	Oak with a long narrow willow leaf, vulgo willow oak.
Quercus alba foliis superne latioribus, opposite sinuatis, sinubus angulisque obtusis.	White oak, with the upper leaves broad, oppositely sinuated, the sinusses having obtuse angles, or the true white oak.
Ilex Floridana, foliis dentatis, baccis rubris.	Floridan holly, with indented leaves and red berries.
Acer foliis compositis, floribus racemosis.	Maple, with composite leaves and the flowers in bunches.
Acer foliis quinque partito palmatis acuminato dentatis.	Maple, with a palmated leaf of five parts sharply indented.
Acer foliis quinquelobis subdentatis, subtusglaucis pedunculis simplissimis aggregatis.	Maple, with a five lobed leaf faintly indented. their lower part of a blue cast, with simply aggregate flower stalks.
Fraxinus floridana, foliis angustioribus utrinque aecuminatis pendulis.	Floridan ash, with narrow hanging leaves on both ends pointed.

Nyssa foliis latis acuminatis et dentatis fructu aeleagni majore.
Tupelo, with broad pointed and indented leaves, with a fruit like the largest wild olive.

Populus alba majoribus foliis subcordatis.
Great white popular, with almost heart shaped leaves.

Populus nigra folio maximo, gemmis balsamum odoratissimum fundentibus.
Black popular, with the largest leaves, whose buds exude an odoriferous gum or balsam.

Platanus occidentalis foliis lobatis.
Western plantane, with lobated leaves (vulgo) button wood, water beech, or sycamore.

Salix folio angustissimo, longissimo subtus albo.
Willow, with narrow long leaves, being white below.

Bignonia foliis simplicibus cordatis, flore sordide albo, intus maculis caeruleis et purpureis irregulariter adspersis; silique longissima et angustissima.
Bignonia, or trumpet flower, with single h e a r t shaped leaves, flowers of a dirty white, through whose inside blue and purple spots are irregularly scattered, having a long and narrow seed pod (volgo) catalpa.

Bignonia, fraxini foliis.
Bignonia, with an ash leaf.

Laurus foliis acuminatis, baccis caerurleis; pedicellis longis rubris infidentibus.
Laurel, with a pointed leaf, blue berries sitting on long footstalks.

Crataegus fructu parvo rubro.
Hawthorn, with a small red fruit.

Genista capsulo aromatico.
Broom, with an aromatic seed pod.

Vitis nigra, vulpina dicta.
Black vine, called fox grape.

Vitis foliis api, uva corymbosa purpura minore.
Parsley leaved vine, with small purple grapes in a corymbus.

Vitis vinifera silvestris.
Wild wine vine.

Betula nigra foliis rhombeis ovatis acuminatis duplicato serratis.
Black birch; with ovate rhomboid leaves, being doubly serrated.

Coriaria foliis gladiatis serratis (seu) nicotiana Indiorum.
Shumac, with serrated sword like leaves, or Savages tobacco.

Rhus vernix (seu) toxicodendron foliis alatis fructu rhomboide.
Shumac or poison tree, with winged leaves, and a rhomboidal fruit.

Juglans alba aquatica, cortice glabro, arbor humilis; fructu amaro.
White swamp hickory, with a smooth bark being a dwarf tree and a bitter fruit.

Sambucus racemosa acinis nigris, caula herbacea.
Elder, with bunches of black berries, and an herbaceous stalk.

Sambucus rymis quinquepartitis foliis subpinnatis.
Elder, with the cyma of five parts and imperfectly winged leaves.

Magnolia, glauca lauri folio subtus albicante.
Magnolia, with a green laurel leaf whitish below.

Fagus foliis ovatis obsolete serratis; fructu triangulo.
Beech, with almost egg shaped slightly serrated leaves and a triangular fruit.

Myrica (seu) myrtus (brabantica similis) floridana, baccifera, baccis sessilis; fructu cerifero.
Florida berry-bearing myrtle, the berries squat; and yielding wax.

Canna foliis enervibus.
Reed, with very week leaves.

Gleditsia spinosa, spinis triplicibus axillaribus, capsula ovali, unicum semen claudente.
Locust, with triple axillary spines, an oval seed pod inclosing a single seed.

Salix folio angustissimo serrato glabro, petiolis dentatis glandu losis.
Willow, with very narrow smooth serrated leaves, the stalks dentated and full of glandules.

The back or inland swamps answer in situation to what are called the meadows or savannahs (among the pine lands) their soil being rich, occasions them to bear trees. The true back swamps, that are in wet seasons full of standing water, bear scarcely any other tree, than a variety of that species of Nyssa distinguished by Botanists by the name of *Nyssa foliis latis acuminatis non dentatis fructu aeleagni minore, pedunculis multiflore,* vulgarly called bottle arsed tupelo; the continuance of water on this kind of ground, is the reason why scarce any undergrowth is found here. There are swamps also called back

Avena aquatica Sylvestris

swamps, but they are either at the head of some stream, or have more or less water running through them; these are generally easy to drain. I would have confined my description of back swamps to the first or standing ones, and ranked the last (which I think might properly be done) among the river swamps, but I was apprehensive, that it might have displeased some person, who entertains the more established opinion; these last described often are found meer cypress swamps, in that case, they are almost impassable, by reason of the cypress spurs, even when dry, and for horses, they are extremely dangerous, as they often get staked on those spurs. This vegetable monster I shall hereafter describe; I do not remember to have ever seen it mentioned any where; when this kind of swamp is not over grown with cypress alone, its product is the same as that of the river swamps above mentioned, and in that case the soil is certainly good; these last when properly drained, are the best land for the cultivation of hemp.

The marshes are next to be considered, they are of four kinds, two in the salt, and two in the fresh water; they are either soft or hard, the soft marshes consisting of a very wet clay or mud, are as yet of no use, without a very great expence to drain them; the hard ones are made up of a kind of marly clay, which in dry seasons is almost burned up, true it is they afford a pasture sufficient to keep any gramenivorous animals in good order; but the milk and flesh of them in seasons, when the cattle near the sea side cannot find any other food, and consequently feed on this alone, are of so horrible a taste, that no stranger to the country can make use of them. Hard marshes in general are such, whose soil has too much solidity, for the water to disunite its particles by penetrating them; the soft marshes are those, whose spungy nature allows the water easily to penetrate them; I have seen of both kinds on Turtle River, about twenty miles up, in which, at about eight or ten feet below the surface, there are numbers of cypress and other stumps remaining, but chiefly cypress, and many of the fallen trees crossing each other; this is only to be seen at low water, and to the height above named; these trees are covered with a rich nitrous muddy soil; but I beg leave to expect, that better Naturalists may explain this extraordinary appearance; I believe them ruins of ancient forests on which the sea has encroached.*

The marshes on fresh water are in every respect similar to those on the salt, except, that they are not impregnated with the saline particles, of which the first are very replete; therefore the hard ones, with little trouble, are adapted to cultivation; the soft ones cost a considerable deal more of expence, to render them fit to answer this purpose, but when so drained as to answer the end, they certainly are by no means inferior to any land in this country; in the lower part of these marshes grows a kind of hitherto undescribed grain, of which the western Indians make a great use for bread, I never could see it in blossom, therefore, cannot describe it, but joined to this is a figure of it, nearly equal in size to one eighth of the com-

---
*The whole appearance of this river seems to indicate such an ancient and unrecorded hurricane on this part of the coast.

mon growth of the plant when in perfection, it is known by the name of wild oats.

This kind of land produces rice very willingly, but if sufficently made dry, always proves very good for corn, indigo and hemp; I have seen at Mr. Brewington's plantation, about three miles below savannah in Georgia, very good corn and rice together, with the two kinds of melons, and cucumbers in great perfection on this species of soil.

I shall next describe the bay and cypress galls; these intersect the pine lands, and are seldom of any breadth; the bay galls are properly water courses, covered with a spungy earth mixed with a kind of matted vegetable fibres; they are so very unstable, as to shake for a great extent round a person, who, standing on some part thereof, moves himself slightly up and down; they often prove fatal to cattle, and sometimes I have been detained for above an hour at the narrowest passes of them, they being so dangerous to cross, that frequently a horse plunges in, so as to leave only his head in sight; their natural produce is a stately tree called loblolly bay*, and many different vines, briars, thorny withs, and on their edges a species of red or summer cane, which together combine to make this ground impenetrable, as if nature had thus intended to prevent the destruction of cattle in these dismal bogs, which would be particularly fatal to many of them in spring, when the early produce of grass and green leaves in these galls, might entice them into this danger, was not such a natural obstacle in their way; as these have generally vent, they are sometimes drained, and rice planted therein, which for one or two years thrives there tolerably, but this ground is so replete with vitriolic principles, that the water standing in them in impregnated with acid, insomuch, that I have tasted it sour enough to have persuaded a person, unacquainted with this circumstance, that it was an equal portion of vinegar and water mixed together, therefore it requires to lay open at least one year before it will bear any thing, and they generally, by laying open four or five years without any other draining, becomes quite dry, and might be advantageously used for pasture ground.

The cypress galls differ from these, in being a firm sandy soil, in having no vitriolic taste in the water, and very seldom vent; I never knew these made use of for the purpose of planting, and the cypress they produce, is a dwarf kind, not fit for use, being very much twisted and often hollow, there is no undergrowth here, but in dry seasons

---
*Hyperium, seu Gardenia Lasianthus.

some tolerable grass. Through all the above species of land we find a distribution of very fine clay, fit for manufacturing; the finest I ever saw is at the village on Mobile Bay, where I have seen the inhabitants, in imitation of the Savages, have several rough made vessels thereof; there is also a great variety of nitrous and bituminous earths, fossills, marles, boles, magnetic and other iron ore, lead, coal, chalk, slate, free stone, chrystals, and white topazes, these last in the beds of rivers; ambergris is sometimes found; one Stirrup a few years ago found a piece of a very enormous size on one of the keys; there is also much of a natural pitch or *asphalthus*, vulgarly called *mungiac*, thrown up by the sea: The uplands also afford a metallic substance appearing like musket bullets, which on being thrown into the fire go off in smoak with a very sulphureous stench.

The water in this country is very various as to taste, quality and use, there are salt, brackish, nitrous sulphureous, and good fresh springs in most parts of this country, as well as salt and fresh lakes, lagoons and rivers, the rivers also vary in many respects, and so does the sea as well in the colour, and clearness of the water as in its degree of saltness; the water of St. Mary's and Nassau, and all the brooks that run into them is very good, wholesome, and well tasted, the colour in the rivers is dark, as in all the American rivers of the southern district; St. John's is a curiosity among rivers indeed, this rises at a small distance from the lagoon called Indian river, somewhere in or near the latitude of 27, perhaps out of the lake Mayacco, which I have reason to believe really exists, and is the head of the river St. Lucia, as I am told by a credible Spanish hunter, who had been carried there by way of this last river; from its origin it runs through wide extended plains and marshes, till near the latitude 28, where it approaches the lagoon much, it then continues its course with a considerable current northward, and glides thro' five great lakes, of which the last, called Lake George, is by much the most considerable; in this last lake is about eight feet water, it is twenty miles long and about eleven or twelve wide; all these lakes and the river in general is very pleasant; endless orange groves are found here, and indeed on every part of the river; below these the river grows wider, loses it current, and has in some places none, in others a retrograde one, when yet lower down it is again in its true direction; the banks of this river are very poor land and exhibit in a number of places sad monuments of the folly and extravagant ideas of the first European adventurers and schemers, and the villany of their managers; the tide does not effect this river very far up. In many places high up this river are found some extraordinary springs, which at a small distance from the river on both sides, rush or boil out of the earth, at once becoming navigable for boats, and from twenty five to forty yards wide, their course is seldom half a mile before they meet the river; their water is (contrary to that of the river) clear, so as to admit of the seeing a small piece of money at the depth of ten feet or more; they smell strong of sulphur, and whatever is thrown in them becomes soon incrusted with a white fungous matter; their taste is bituminous, very disagreeable, and they in my opinion cause the green cloudings we see on the surface of the water of this river, and make it putrid, and so unwholesome as experience has taught us it is. I have no sufficient ground to decide upon another circumstance, which I am told, viz: That when rice is overflown with this river water, it kills it; above these springs the water of the river is good: This river is from one and half to three miles wide, except at the house of Mr. Rolle, who has here made an odd attempt towards settling and making an estate in as complete a sandy desart as can be found; just above this, it is full of islands, exhibiting every where a very romantic appearance; there is a fine piece of water, called Dun's Lake, this is about nine miles from the river, eastward from this place, this empties itself by a stream into the river; another called the Doctor's Lake, is on the west side, about sixty miles from the mouth, we see a variety of aquatic plants floating thereon. In my journey by land from the Bay of Tampe across the Peninsula to St. Augustine, I crossed twenty three miles from east to west of miserable barren sand hills, the grain of the sand is very small and ferrugineous; these hills rise a considerable height; on them is some growth of very small pines, and a very humble kind of oak grows so thick, that with the addition of some wythes and other plants, to me utterly unknown, they render it absolutely impenetrable. In this Ridge, which, as far as I can learn, extends from North to South, between the rivers St. John and *Ocklaw-wawhaw*, for about an hundred and fifty miles, having no where any water in its whole extent, and I am told, that where we crossed it, is its narrowest place; my Indian guide had the precaution to carry water for ourselves and horses, which proved very serviceable as it was a very hot day, no growth of trees to shade us, and such a burning sand for the sun to reflect on; I leave the reader to judge what we suffered, though it was but a short distance over, both ourselves and beasts often experienced the necessity of carrying water;

what must travelling over this place be in a hot day, where it is forty or more miles wide?

Before I leave St. John's River, I must not forget the river running from south to north, called Pablo: This originates at a small distance from St. Marks or North River, and empties into St. Johns at a small distance from the mouth. The water of this river is good, so is the land on it; and it is thought that a communication with St. Mark's or the North River might be effected without much difficulty: this would open an inland navigation by canoes or boats, all the way from Carolina to near the Musketo.

The river St. Mary although it is said to originate in the *Aekanphanaekin* swamp has a current of fine clear and wholesome water supplied from the pine lands through which it flows, with many fine springs, runs, and rivulets of very clear water, Nassau has also the same blessings, but doth not spring far distant from the sea. On Amelia Island near the sea, is a very good spring, which makes a fine stream for some miles, dividing the island almost into two; but below the spring its water is not commendable. On the beach between St. John's and St. Augustine, at or near a place called the Horseguards, there are three good springs running into the sea, and in every part where the beach is clear sand, water is obtained by digging. About four miles north of St. Augustine rises St. Sebastian's Creek, being a good fresh spring, it soon joins a creek in salt marshes, and at a small distance from town it becomes very large and deep; it empties into St. Anastasia's Sound two miles south of St. Augustine, making a Peninsula of this territory nearly in form of a crescent; three miles farther south is the mouth of the river St. Nicholas not very considerable; St. Cecilia in the same sound, the North west, south of the Matanca and Penon; the Tomoke and Spruce Creek in the Musketo Lagoon, and in short every river and creek in the country except those above named bituminous ones, are excellent wholesome water; thus much I suppose will suffice as to the nature and quality of the water: All the rivers and springs in West Florida are good.

The aborigines of the country come most naturally under our next consideration, no one is ignorant, that the epithet of indians is given to those people, though no doubt, the French name of savage is a much more proper one, as the manners of the red men are in every respect such as betray that disposition, and shew the savage thro' the best wrought veil of civilization; we might call them Americans, as the inhabitants of the old world are each distinguished by a name expressive of, or relative to the quarters, from which they respectively originate, but this would be confounding them with the other natives, as well white as black, which I think by no means reasonable.

Oldmixon with all his failings is undoubtedly the only writer, who speaks with truth and pertinence on this subject; all other English, French, and Spanish authors, which have fallen in my way (and they are not a few) have made of this story a confused heap of nonsense and falshood; I shall relate what I know and have found from real experience among four or five nations, and as I can vouch for the similarity, that will be found among them, I believe my reader will be of my opinion, that from one end of America to the other, the red people are the same nation and draw their origin from a different source, than either Europeans, Chinese, Negroes, Moors, Indians, or any other different species of the human genus, of which I think there are many species, as well as among most other animals, and that they are not a variety occasioned by a commixture of any of the above species must also appear.

The above account will perhaps raise a conjecture that I believe the red men are not come from the westward out of the east of Asia; I do not believe it, I am firmly of opinion, that God created an original man and woman in this part of the globe, of different species from any in the other parts, and if per chance in the Russian dominions, there are a people of similar make and manners, is it not more natural to think they were colonies from the numerous nations on the continent of America, than to imagine, that from the small comparative number of those Russian subjects, such a vast country should have been so numerously peopled, and by what we know from the geographical discoveries that have been made within this century, it was undoubtedly easier for these people to have crossed out of America into Asia, than it was for the white people we find in Labrador to come from Lapland, yet who will deny that a Laplander and *Eskimaux* are of the same original stock; add to this, that I have both sacred and prophane history, besides daily experience on my side to prove, that population, as well as all other things we find in nature, have always moved from the eastward, and still continue so to do, why then should we insist on one part of this system to move in a retrograde way, when for a further proof we find, by what we learn from the savages that have been far to the northwest, that some white people answering to the Japanese, sometimes come on that coast, but do not stay, nor have ever attempted colonization.

But alas! what a people do we find them, a people not only rude and uncultivated, but incapable of civilization: a people that would think themselves degraded in the lowest degree, were they to imitate us in any respect whatever, and that look down on us and all our manners with the highest contempt: and of whom experience has taught us, that on the least opportunity they will return like the dog to his vomit. See there the boasted, the admired state of nature, in which these brutes enjoy and pass their time here! How justly did the above named author exclaim:—"Let the learned say all the fine things that wit, eloquence, and art can inspire them with, of the simplicity of pure nature, and its beauty and innocence, the savage wretches of America are an instance, that this innocence is a downright stupidity, and this pretended beauty a deformity, which puts man, the Lord of the creation, on an equal foot (yea below) the brute beasts of the fields and forests."

In describing a people an historian is obliged to speak as they generally are; Dr. Blackwell, drawing consequences from what he imagined a state of nature to be (and what I believe it may have been) among most of the nations on the old continent says:

"In the infancy of states, men generally resemble the publick constitution, they have only that turn, which the rough culture of accidents, perhaps dismal enough, through which they have passed, could give them, they are ignorant and undesigning, governed by fear, and superstition its companion; there is a vast void in their minds, they know not what will happen, nor according to what tenor things will take their course, every new object finds them unprepared, and they gaze and stare like infants taking in their first ideas of light."

How opposite is the savage, he is cunning and designing, knows no fear, nor has he any idea of religion to make him superstitious, on the contrary, the pretended conjurer who lives with him, runs a perpetual risque of his life, he has no void in his mind, but is very deliberate and careful in his mischief and cruelty; the study of what may occur in the next war or hunting season always employs him, he can ever so plan his schemes, as to be certain of his future safety, and success; without this last he neither undertakes nor risks any thing; new objects are but a momentary surprize to him, and his gazing and staring always end in sovereign contempt.

But to demonstrate more clearly the contrast between a savage and the people of the old continent, I beg leave to observe a few things, which have fallen under my inspection and consideration, which I apprehend will be thought pertinent to this purpose: A savage comes into this world with all the possibility of opposition to us; his mother on her approaching labour, retires from all company, aid, or assistance, into some lonesome wood, and there without perceptible pain or inconvenience, disburthens herself, goes into cold water to cleanse herself and her offspring, and returns to her daily vocation.

A savage has the most determined resolution against labouring or tilling the ground, the slave his wife must do that, and a boy of seven or eight years old is ashamed to be seen in his mother's company. No greater disgrace can be thrown on a man than calling him by the odious epithet of Woman; what other nation do we find so absolutely neglecting agriculture? What people are ashamed to be seen eating or drinking in company with the fairer sex?

Our women carry their children with their faces towards their own, a she savage puts the back of hers towards her own back. When we make fire, we pile the fuel parrallel to each other; the savage puts his wood in a circular form, lights the central ends, and by the help of one of the sticks, which he shoves always to the center, he keeps it in.

We make war in an open brave way, a savage by hiding himself surprizes; our prisoners are sure of life, the prisoner of a savage is sure to die by cruel tortures. When we take a sweat, we keep ourselves warm with the utmost care; a savage with his open pores, plunges head long into the almost frozen river, or into a hole on the ice if quite frozen.

A savage man discharges his urine in a sitting posture, and a savage woman standing, I need not tell how opposite this is to our common practice.

A savage never eats salted meats, nor boils any thing in salted water, though he has salt in abundance, but when he has been a long while from salt and then gets it, he will frequently eat a pound of it without any thing else. A savage either buries none of his dead, or if he does he puts the body in a sitting or standing posture; in a word, if they had always studied to be in contrast with us, they could not be more so, than nature has made them.

O Deus! homines et homines creavisti.

All savges, with whom I have been acquainted, are, generally speaking, well made, of a good stature, and neatly limbed; crooked, lame, or otherwise deformed persons are seldom or never seen among these people; and if per chance we find a savage labouring under these misfortunes, we always find them accidental, never natural. There colour resembles that of cinnamon, with a copperish cast; they are born white, but retain that hue a very short time; their hair is lank, strong, black, and long; they prevent the growth of what little beard nature has given them, by plucking it out by the roots; they never suffer any hair to grow on any part of the body except the head, their eyes are black, lively and piercing, and they are blessed with an amazing faculty of discovering objects at a vast distance; their teeth are very good, and to the last they retain them, being never subject to the tooth ache; they are strong and active, patient in hunger and the fatigue of hunting and journeying, but impatient and incapable of bearing labour, they are incredibly swift of foot; their discourse is generally of war, hunting, or indecency; their women are handsome, well made, only wanting the colour and cleanliness of our ladies, to make them appear lovely in every eye; their strength is great, and they labour hard, carrying very heavy burdens a great distance; they are lascivious, and have no idea of chastity in a girl, but in married women, incontinence is severely punished; a savage never forgives that crime. They are capable of an attachment, rather than a friendship; addicted to lying in a high degree; their seeming candour and simplicity is an effect of dissimulation; they know how to save appearances, and will always find ways to cover their knavish, thieving tricks; their notions of faith and honour are such as to make them violate their word of promise, even when they are in treaty, unless compelled to be true by fear or force. They are brutal and have not the most distant idea of decorum; without taste they are terrible drunkards, in which last state, there is no villany nor crime, they will not commit, and when they recover their senses, throw the blame on the liquor, holding themselves entirely excused; no religion of any kind can we trace among them. Possessed of indifference and want of sentiment, they drag themselves through a kind of life, which would make us pass our days very irksomely, and tire us in a short time with the disagreeable similitude of our hours; in a word, they have nothing in their way of life to tempt a man of the least reflection, to envy them their miserable state of nature.

But with all these bad qualities, they have one virtue, which is hospitality, and this they carry to excess; a savage will share his last ounce of meat with a visitant stranger. What travellers have related of their giving their daughters to transient persons, is not true, and it it not till after some acquaintance, that they will give a white man, what they call a wife, unless he chooses an abandoned prostitute, which are here to be found as well as elsewhere. Among the nations, which I have frequented, a young stranger (led away by the notion of this traveller's story) that would attempt, on his first acquaintance, to ask a man for his daughter, might pay dear for it. The Chactaws, who have a greater idea of a *meum* and *tuum* than any of their neighbours, are not so hospitable as they; but although they will on this principle require pay for even the least morsel, they set before one, yet the same idea of property has taken off much of their savage dispositions in many other cases.

I cannot think it foreign to my subject to mention, before I proceed to the particular description of each nation, something farther about the origin of the native Americans: Great pains have been taken to prove these people descended from the *Phaenicians*; among others I find an anonimous Gentleman, who gravely tells us, that *Phaenicians* or *Erythreans* in Greek, carry the same sense as *Esau* or *Edom* in Hebrew, both meaning red, hence that nation was red, so are the savages, ergo, they are the same people; as ridiculous as this argument is, the pains taken to shew that the extensive navigation of the *Phaenicians* amounts to a proof of their having crossed the Atlantic to America, is no less void of sense, although far fetched annotations are made use of to prove their naval power, by *Cambyses* being obliged to renounce his design on *Carthage*, by reason of the *Phaenicians* refusing to put to sea, yet neither that, nor the proof of their having built *Leptis, Utica, Hippo, Adrumetum*, on the African continent, nor *Calis*, and *Tartessus* in Europe, or their colonizations in *Iberia*, and *Lybia*, nor all the power ascribed to them by *Curtius*, seems in the least to indicate any transit to this western world.

Equally absurd are the reveries of *Comtaeus*, of those who construe the island discovered by the Phaenicians (which *Diodorus Siculus* mentions L. 6 C. 7, and the passages of *Pliny* L. 5 C. 1 and *Pomponius Mela* L. 1 C. 4) likewise the common wealth of the alegorical dialogue between *Midas and Silenus* by *Theopompus*, and the quotation of *Pausanias* concerning the discovery of the *Satyrides* by *Euphemius* of *Caria*) into America; also *Bossu*, Father *Laffiteau, du Pratz*, and others, as the refutation of these would be uninteresting and therefore tedious, I shall content myself with having only mentioned them, and proceed to take some notice of the more general hypothesis; *that the savages are the dispersions of the ten lost tribes of Israel.*

We are told, that the Americans agree in many of their customs with the Jews; but in which? We see not the smallest similarity in any of their ways, unless the separation of women among the Chicasaws, at the time of their *Catamenia;* but not to urge how naturally this fashion might point itself out, and its being confined to one small tribe only, let us argue a little on the jewish grand and characteristic ceremony of circumcision, of which the boldest genealogist in this way, cannot find any the least mark, nor traces among those people. How have the jews neglected to introduce this sign of the covenant here, when they have made it obtain among so many of the Easterns where they lived even as slaves and exiles; and they always thought their salvation dependent thereon? How many tribes in Africa still retain that custom, though without any other of the Isralitish rites, which they have forgot? but where can we discover even the most distant appearance of this ceremony in America?

Just so is it with that pretended migration to *Assareth,* as mentioned in the fourth book of *Esdras,* in the thirteenth chapter; but without insisting on the apocryphal quality of this evidence, I would only ask, why this *Assareth* should be more America, than any other far distant country? And how they found their way by the *Euphrates* to this continent? Further it is evident, that *St. Jerome* (who flourished under *Theodosius* in the latter end of the fourth century, lived in *Asia,* and held a close correspondence with the Jews, for the sake of learning their language) plainly tells us in his *Ezech*: L. 5, and *Jerem*: L. 6, that those very ten tribes, were then groaning under the severe yoke of slavery, in the cities of *Media* and *Persia;* this I judge to be proof enough to invalidate this journey to *Assareth,* and rank it along with the fables of the loss of the whole law, the confinement of souls in subterraneous apartments, and with the childish tales of *Behemoth* and *Leviathan* in the same fourth book of *Esdras.*

Yet, as a farther proof of the falshood of this American expedition, by the ten tribes of *Israel,* I would beg leave to ask my readers, whether they can imagine, that so prodigious a migration could possibly happen at a time of the most desperate wars, and desolations in the very countries through which they must necessarily pass, for it is evident, that since *St. Jerome's* time, although the *Persians* had again regained their empire, yet for several centuries together, they were constantly warring with the *Romans, Indians,* and other Eastern nations, and lastly with the *Saracens.* 'Till the famous Tartarian expedition under *Zengis Chan,* this whole country swam in the blood of its natives, and before this period, the Grecian wars, the conquests of *Alexander,* with the divisions of his captains at his death, furnished as little opportunity for this journey, which also must have been strangely overlooked, by such a number of historians of credit, as flourished during the above space of time. I could with as much force of argument invalidate *Emanuel de Moras* (a Spaniard;) who, to prove that America is peopled from *Carthage,* and by the Jews, makes numerous far fetched comparisons, of the manners of these nations and the *Brazilians;* nor can I think, that after this, any of the favourers of this Jewish ancestry will insist, that the natives of America are descended from the other two tribes, since they were not only not allowed to see the land of their fathers, even in quality of travellers, but on pain of death, were they forbid to assemble in large companies, which it was however necessary they should do for such a migration; but I am afraid my reader, as well as myself, is by this time tired of so much argumentation against so silly an Hypothesis.

No less idle is the argument used by Grotius concerning the Mexicans telling the Spaniards, that they came out of the North, where he thinks to find the Norwegian descent in the city *Norumbega,* but who can find in North America, traces of any ancient city at all? That the Americans divide their time by nights (or sleeps, as they stile them) is true, also their dipping new born infants in cold water, their going naked, and their eating separate from the women, with many other customs of the savages, similar to the ancient manners of the Germans, but this only proves to me, that in the times of *Tacitus* and the other *Roman* historians, the *Teutonick* nations were in as savage a state of nature, as we now see the Americans; a sameness in colour, and stature, as well as, or rather than similarity of Languages is to me a requisite proof, and for that reason I believe, that the *Eskimaux* came from *Lapland,* but by what accident, we are ignorant of, and so we might guess at some eventual passage of the red savages, were we able to trace any similarity between them and any nation of the old world, but we see none; yet all the islands of America, even *New Zealand,* are peopled by the same race as this vast continent; therefore I will rather suppose, that the savages, have sent out colonies, than that they are a colony from any other part of the earth. The same

Gentleman endeavours to fetch the *Peruvians* from *China,* but with as little foundation in truth. Mr. Powell's story of Maddock the Welshman's voyage, cannot be thought fit, to ascribe a Welsh origin to the savages, because they could by no means either grow to such numbers nor degenerate so much in colour, as they must have done, in two centuries; yet even this has been made use of by a certain Dutch writer, whose name I do not now recollect; and how unluckily does *Bossu* here bring forth the black headed *Pinguin,* to prove that the Welsh word *Pen-gwin,* was given to this bird for a name, by Maddock's followers, because *Pen* is head, and *gwyn* is white, and Mr. Foster in his note on that passage, is as unlucky in saying *Penguin* is Spanish for a fat bird.

I find a more powerful argument used by those who bring the savages from *Tartary;* because I do not at all doubt, from my judgment of geography, that *America* nearly joins *Asia;* but the expedition of *Hoccota,* the son of *Canguista,* first King of the Tartars, as mentioned by *Michalon Lituanus,* in his Ennead 9, L. 6, which has likewise been construed to have reached America, is also too late to allow them time to become so numerous a people; because this man lived about the year 1240; but although I believe a possibility of the passage, yet I cannot find, that there is any similiarity in the persons of the North Eastern Tartars, and the American savages, and much less in their manners; and if there is between those of *Kamschatka* and America, I still think, that the first are more naturally to be supposed a small colony from the latter, than that the latter prodigious numerous people are sprung from so small a tribe as the *Kamschatkans;* every one of the red men over the whole continent, and those who still remain in a few of the islands, is exactly similar in person to his neighbour, and their numbers are to this day vastly great, notwithstanding the cruel depredations of the several European nations.

The contemptible light, in which *Bossu* endeavours to place the number of the savages, is well confuted by Mr. Hutchin's account of the nations near our settlements, who having mutually destroyed each other, and been destroyed by the sword, and liquid fire of the Europeans, at a shocking rate, yet amount by that Gentleman's moderate calculation, even at this day, to near three hundred thousand souls; therefore it is not their small number, but the vast extent of this continent, that causes their nomadic life.

But to treat more particularly against this notion, of the American descent from the Tartars, who can suppose that their darling animal, as well for food, labour, and war, as other uses, I mean the horse, should not have been brought along with the colony? Who knows not, that the Tartars use most, and can least spare this noble creature? Does not all their martial power consist in horsemen? Does not their very maintenance depend on horses? Have we not sufficient proof, that many Tartarian Lords keep hundreds of horses, and even stable them? None are so poor as not to have two or three; when they shift their abode, do not horses carry their tents and provisions? When all other food fails them, garlick and mare's milk is their resource; the very blood out of the veins of these animals supplies the place of an agreeable drink; what stream so rapid which they do not cross on their horses backs, or by holding their manes, and guiding them with a twig, while they swim at their side? making a small raft of their baggage, they tie it to the tail, and thus cross every river in their way; a horse's skin makes a boat to cross a very wide river, lake, or even an arm of the sea; the supposed narrow division between Asia and America, near Cape *Tchukshi*, could then by no means, be an obstacle to the transportation of these animals; the division of the several places, named by *Bossu*, through earthquakes, is absurdly introduced, to prove that this migration happened before such a convulsion of nature, which prevented the return of any of these emigrants. A Tartarian army appears larger than it is, on account of each man having two or three horses; and notwithstanding their horses die by the sword, or are killed for food, yet many thousands are yearly delivered to *Russia*, by the European Tartars only; this much premised, how had the savages no horses on the arrival of the Spaniards? Had the Tartars inhabited a country near America, the very horses would have increased so as to come over to this continent; how many instances do we find even among the settled provinces of America, of horses going wild! but the greatest proof, that this must have infallibly happened, is to be found in South America, where now many millions are found wild, in the great plains south of Brazil, especially near *Rio de la Plata;* yet these take their origin from the Northern America, and have crossed vast mountains; how it came it then, that a dozen or two of horsemen routed myriads of Americans? only one objection remains to be confuted; there may be an arm of the sea between Asia and America, and this colony came by water; but can it even then be supposed, that these people would have forgot their favourite horses? or will any say that if the horses could come by land, yet every climate does not maintain them; one summer would suffice for the journey: A horse when travelling alone, is experienced to go very fast forward on his journey, and we learn also from the same great teacher (experience) that thousands of horses in North America, even breeding mares, live in the woods in deep snows, through hard winters without human assistance; but no, these Tartars could not bring horses, nor other useful animals, and yet a colony of Panthers, Lynxes, Wolves, Foxes, Otters, and other destructive animals must be supposed to have found their way as well as these emigrants. Here language is tortured into a proof; a certain *Abraham a Mylis* has found out, that in the empire of *Cathay,* which lays nearest America, they speak the *Teutonic* dialect, and this same empire being divided, into seven provinces, the Eastermost is called *Tendhuk,* very like to *het eynde des hoeks,* or the end of the point, and *Anyan* is very like, *Aangang,* or passage (a passage to be sure out of Tartary, to America;) again beyond *Anyan* is the vast country of *Bergo,* this certainly is very much like *Bergen,* or sheltering, because the *Scythians* leaving their native country, sheltered themselves here; fine Etymology! noble Geography! (*risum teneatis amici*!)

The passage from Asia to America, is by no means proved by the pretended Elephants bones found in the swamp near Ohio, since by later observations, and examinations of the bones, by men of more knowledge than the meer vulgar, they appear never to have belonged to elephants, but more probably are the remains of some *Hippopotami,* especially as near that river, the Indians, often at this day, discover large foot steps, and hear great bellowings, both proceeding from animals they never see themselves.

I think if Mr. *Bossu* had obliged us with the knowledge of the particular nation that calls the *Ginseng,** *Gareloguen,* he would have thrown more light on his far fetched notion of the parallel between the Tartars and Savages; and since he has himself invalidated the pretended etymology, I would beg leave to deal with his similar significations in the same manner, for why should not the same idea strike *Bossu's* nation and the *Mantcheoux* Tartars, upon sight of the *Ginseng,* and make both compare it to men's thighs? But as but one nation among so many has accepted this idea, it cannot be allowed,

---

**Panax Quinquefolium.*

even as the most distant proof; the fable of the *Escaaniba* by Bossu also, is too ridiculous and absurd to merit confutation.

Nothing now remains, but a few words about the migration of the *Chinese* into *Peru,* from whence some suppose America is peopled, but how different the colour and figure of the people! and where do we find any *Chinese* finesse among the savages? besides, what Geographer does not know that the *Chinese* might more easily, have come over to the eastern shore of America, across the Atlantic, than over the wilder, wider, and more stormy South Sea, to land on *Peru's* or *Chili's* unhospitable shore, where our large European vessels can scarcely live, much less a *Chinese* junk; nor did the *Peruvians* know any thing of large Vessels till the arrival of the Spaniards; strange indeed, that their naval knowledge, and mother country, should be thus forgotten! the more so, as it is much easier, to go from America to *China,* than the contrary, by reason of the constant easterly winds reigning here, which also make it more improbable that the *Chinese* vessels came here by chance, or stress of weather; nor could they in that case, but by chance, be provided with the provisions necessary for so long a voyage.

Methinks I hear some critical person say, whence then came the savages? That indeed, is a difficult question to answer; it is easier to confute the different opinions, than to say any thing with certainty on this head, more especially as these people want all manner of record; yet a tolerable guess may be made to extricate us out of this difficult labyrinth.

Without doubt *Moses's* account of the Creation is true, but why should this Historian's books, in this one thing, be taken so universally, when he evidently has confined himself to a kind of chronicle concerning one small part of the earth, and in this to one nation only; this account therefore must be understood with the same limitation as many others in holy writ are generally allowed to be; I think therefore that (as mentioned before) we do not at all derogate from God's greatness, nor in any ways dishonour the sacred evidence given us by his servants, when we think, that there were as many *Adams* and *Eves* (every body knows these names to have an allegorical sense) as we find different species of the human genus; is this not a more natural way, agreeing more with the proceedings of a God of order, than the silly suppositions that the variety is an effect of chance, much less a consequence of curses? the more so, as neither has any foundation in God's holy word; why must we think that the curses of a head of a family, should effect each race with a peculiar set of features, and shape of body? Or one tribe with a red, a second with a sallow, a third with a yellow, and a fourth with a black colour, &c? Besides we read of but one curse of this kind; they are different species then. Anatomy has taught us, that the bone of a Negroe's skull, is always black, that besides the *Tunics* of which our skins are composed, they have an additional one, consisting of numerous vesicles, filled with a black ink-like humour; see here then two characteristicks, besides the blackness of the skin, whereby this sable race is distinguished; may not experience teach us, that when the other species are more carefully examined, they will all be found to have some such peculiar character of distinction? Why then shall we involve ourselves into numberless, needless difficulties about the origin of these so singular people, so very different from all other tribes on the globe, yet so very similar to each other: Throughout their own continent their wild manners are universally alike; their languages only differ; why then can we not take the more easy way in saying, God has created an original pair here as well as elsewhere: Is not this opinion supported by our finding numberless other kinds of the animal kingdom on this continent peculiar to itself; where (besides in America) do we find the *Bos Americanus?* Where the *Paca,* and *Vecunha* (both American sheep?) where the *Ursus Luscus* or racoon, the *Armadillo,* the *Agouti,* the *Lacerta Americana,* called *Leguana?* the *Ignavo* (an animal called the American Sloth?) where the same kind of *simiae,* and *Vespertilio Cynocephalus?* where the *Warri* or the *Pecairi* (two species of the *Sus?*) The many kinds of *Myrmecophaga,* the *Mus Marsupialis, Mus Scalopes, Mus Palustris hispidus,* of the same kind, as those in America? Where again the *Struthio Americanus,* or the immense variety of the *Psittacus,* peculiar to America? Where the *Toucan,* the *Phaenicopteros* and *Arquato,* with the red *Platea?* Where the beautiful wood duck, and the kinds of *Phasianus* called *Quama* and *Curasoa?* Where the little *Mellisuga,* or the beautiful, and remarkable aquatic *Muscophagus,* called a sun bird? and many more of the quadruped and winged tribes, large and small, too tedious to enumerate? Where again the rattle Snake, and others of the reptile kind? Even among Fishes, the same observation holds; has *Pliny* among all his fabulous animals, described any like these really existing ones?

 Again, whence was America stocked with men in our likeness?

Where were the oxen, the horses, the swine, dogs, cats, or even rats? or where were our poultry to be found on this continent? Yet we see how immensely they have multiplied since their first importation, so that the air and food are certainly well adapted to them; a wise effect of providence, which knew this quarter must one day become inhabited by animals not originally created in it.

What can now be said against this my argument for a separate origin of the savages? Nothing that will amount to an absolute proof of the contrary. But a difficulty arises as to the effect of the deluge; to remove which I must observe, that America, even in its mountains, retains very few if any of the so often quoted marks of this inundation; on the contrary, it is a smooth regularly rising country; and I think, that to believe the flood so far partial, as to have reached the only lands of *Egypt, Palestine, Armenia,* and *Greece,* is by no means an absurd opinion, nor inconsistent with the destruction of the race of men of which Moses treats, nor with the water being fifteen cubits above the highest hill then known, supposed to be *Ararat*; and had it been absolutely universal, yet the *Andes, Cordileras, Aceytas,* and *Santa Martha,* are many hundred cubits higher and consequently, might remain a vast way dry; but grant the deluge universal, who will dispute with me, that the omnipotent Lord of the universe could create a red man after his own image, suffer him to fall as well as a white man, destroy his posterity except a few, and find a way to save a remnant of man and beast, and then for reasons, into which we have no right to enter, any more than the clay has to contend with the potter, leave this race without extending his tender mercies to them, as he did to us, not even sending his apostles into this quarter of the world; and when the christians, so called, arrived here first, to suffer the pretended Missionaries to preach a false, absurd doctrine to those wretches, nay, even to this day to harden their hearts against the light drawn from under a bushel, and placed conspicuously on a candlestick.

However among none of the savages do we find any tradition of such a flood, except among some southern nations, who tell of six persons creeping out of a hole, whereby the earth was peopled; these have been construed into *Noah's* sons and their wives.

The Chactaws have told me of a hole between their nation and the Chicasaws, out of which their whole very numerous nation walked forth at once, without so much as warning any neighbours; I cannot find any relation between this and a deluge: On the *Isthmus* of *Darien* I have indeed been warned by the savages of an approaching flood on a certain day, and when I had stayed till then and saw the event to be nothing, I ridiculed them about it; and they told me that a great while ago they had such an inundation, and that their *Sukies* or conjurors, had foretold a similar one now, but had proved themselves lyars; this is the only hint I could ever trace of any notion of a great flood; and I leave my reader to judge of the weakness or force of this evidence, and of the justness of the opinion advanced, which I offer only because I think that truth should be the object of man's enquiries, and that God has given us the only advantage we have over brutes, in order to spur us on to enquiries into the mysteries of nature.

Characteristick Chucasaw head

I shall now treat of the four most noted nations connected with Florida in particular, beginning with the Chicasaws, they being (although a small tribe) accounted the mother nation on this part of the continent, and their language, universally adopted by most, if not all the western nations. This is the most fierce, cruel, insolent, and haughty people, among the southern nations; they are very intrepid in the wars with their Meridional neighbours, and the French under Messrs. *Artaguette* and *Bienville*, in 1736 experienced their valour (when aided by a few Englishmen of tried courage) and in 1752, and 1753, Messrs. *Benoist* and *Reggio* likewise found them successful to their cost; but with the northern nations it stands otherways; the *Kikapoos, Piancashas* and others are their terror; notwithstanding their boast of scalps from the northward, it has appeared that, except some private murders among nations, who think the insult to come from some other quarter, and some slaves and scalps obtained from the dastardly *Chigtagiks,* they have really not been able so much as to find the *Kikapoos, Piancasha, Wyotani, Shogteys, Musquakey, Otogami* towns; this appeared clearly while I was in the Chicasaw nation in the winter of 1771 and 1772, when one Mr. James (who had been a prisoner with the *Kikapoos,* for the space of three years, and had found means to escape from them) arrived there, and told, that in all that time never one of the Chicasaws appeared, or did mischief among these nations, nor could he learn they ever did; yet it is evident that these come to the very Chicasaw towns to commit depredations, as well as in their hunting grounds, where the Chicasaws always take care to fortify themselves.

It is also observable, that to the north of their towns they never venture any plantations, and to the southward they have many; nay they scarcely venture to get fire wood north of their habitation; in short they have never shown their bravery but against the Chactaws and Creeks, with the French their allies; the Northerns have always been their masters, and the *Arkansas* and *Catawbas* their match; the *Cherokees* have generally had the worst of the wars between them; they have always been staunch allies to the English, but I think there is no very great dependance to be had on them now, for I can (not without reason) affirm, that that alliance was all owing to the French being their irreconcilable, and mortal enemies; now these are out of the way, little reliance is to be had on them; and in the winter of 1771 and 1772, the traders were under daily apprehensions of a quarrel. It is true that those monsters in human form, the very scum and out cast of the earth, are always more prone to savage barbarity than the savages themselves; but it is no less true, that the traders were of as dissolute a life, and of as profligate manners, with as great an inclination for deceit and over reaching during the French time, as they are now; but then the policy of these savages curbed their thoughts of revenge; whereas now they frequently dare to vent threats of a disagreeable nature, which are the more dangerous because those savage politicians are always very much upon their guard when they are treating with our men in office, how they behave themselves in this respect, well knowing that the ancient fame of their faithful alliance is sufficiently rooted in the hearts of the open minded English, to enable them to impose on their credulity.

As an instance of Chicasaw honour or faith, which is indeed equal to that of any nation of American savages whatever; I would mention the tryal that has been made for concluding an union between the *Arkansas, Quappas,* or *Kappas,* and the Chicasaws; the first attempted at least an alliance with the latter so long ago as 1764, but no kind of solemn embassy, or deputation was employed for the effecting this before the summer 1771, and then at this meeting the insolence of the Chicasaws run so high, as to insult one of the *Quapa* deputies by calling him a woman, and spitting rum in his face, which was put up with; but the *Arkansas* never have made an attempt to finish the treaty since, and I dare venture to foretell, that such an union would by no means be for our good, let it happen when it may.

The morals of this nation are more corrupt than those of any of their neighbours; the Chactaws are said to be thieves, but I can assure the reader than the Chicasaws are a thousand times more so; I have had ample proof of it by losing incomparably more in one day at the Chicasaw town than I did in two months going through seventy four Chactaw towns, notwithstanding I had been warned, and was on my guard against the Chicasaws; my razors and a case of instruments, and other trifles of no real use to them, besides every horse I had with me, vanished in one day among these deceitful people. Their discourse is really intolerable, nothing but filth is heard from them; the vanity of being accounted great hunters and warriors has the better of every consideration, and rather than condescend to cultivate the earth (which they think beneath them) they will sit and toy with their women; or if they send them to labour, they play on an aukward kind of flute made of a cane, lolling thus their time away with great indifference, which obliges them yearly to apply for corn and pulse to the Chactaws.

They live nearly in the center of a very large and somewhat uneven savannah, of a diameter of about three miles; this savannah at all times has but a barren look, the earth is very Nitrous, and the savages get their water out of holes or wells dug near the town; in any drought the ground will gape infissures of about six or seven inches wide, and again, two or three days rain will cause an inundation; the water is always nitrous, and this field abounds, with flint, marl, and those kinds of anomilous fossils mistaken for oyster shells, which cannot be burnt into lime; yet this produces a grass of which cattle are so fond as to leave the richest cane brakes for it; and notwithstanding the soil appears barren and burnt up, they thrive to admiration; it also affords a vast, or even immense store of the salubrious *Fragaria*, vulgarly known by the name of wood strawberry.

They have in this field what might be called one town, or rather an assemblage of hutts, of the length of about one mile and a half, and very narrow and irregular; this however, they divide into seven, by the names of *Melattaw* (i.e.) hat and feather, *Chatelaw* (i.e.) copper town, *Chukafalaya* (i.e.) long town, *Hikihaw* (i.e.) stand still, *Chucalissa* (i.e.) great town, *Tuckahaw* (i.e.) a certain weed, and *Ashuck hooma* (i.e) red grass; this was formerly inclosed in palisadoes, and thus well fortified against the attacks of small arms, but now it lays open; a second *Artaguette*, a little more prudent than the first, would now find them an easy prey.

The nearest running water, is about one mile and a half off, to the south of the town, in the edge of the field, but it is of no note; the next is four miles off; and at high times, canoes might come up here out of the river Tombechbé; this place is a ford, which often proves difficult, and on this account is called *Nahoola Inalchubba* (i.e.) the white mens hard labour.

Horses and cattle thrive well in this nation, their breed of the former was once famous, being descended from some Arabian horses brought from Spain to Mexico, but of late they have so mixed them with meaner kinds, as to cause them to degenerate much.

The traders who for fear of causing jealousy by their discourse have formed nick names for all the savage nations, have called these by the whimsical name of the breed; as cunningly suspicious as the savages are, yet I never found that any of them ever took notice of this distinction.

One remarkable thing I cannot omit of this nation: There were in 1771, only two real original Chicasaws left; one of them, who goes by the name of *North West*, scruples not to tell them all very often, that they are of a slave race.

Their grand Chief is called *Opaya Mataha*, and it is said he has killed his man upwards of forty times, for which great feats he has been raised to this nominal dignity, which by all savages is as much regarded, as among us a titular nobleman would be if he should be obliged to be a journeyman taylor for his maintenance.

These savages are the only ones I ever heard of who make their females observe a separation at the time of their *Menses* (some ancient almost extirpated tribes to the northward only excepted, and these used to avoid their own dwelling houses) the women then retire into a small hut set apart for that purpose, of which there are from two to six round each habitation, and by them called moon houses.

The whole tribe are remarkably strong made fellows, but few of their women have regular features, or deserve to be called handsome; these labour vastly hard, either in the field for cultivation of corn, or fetching nuts, fire wood and water, which they chiefly carry on their backs; the two first articles generally two or three miles, and the last often a mile, their burthens would make a stranger, being rather fit for asses than women to carry.

This nation is the most imperious in their carriage towards their women, of any I have met with; they are very jealous of their wives, and adultery in them is punished by the loss of the tip of their nose, which they sometimes cut, but more generally bite off, but this does not deter them, for they are a very salacious race, and the mark is pretty general. They are all good swimmers, notwithstanding they live so far from waters, but they learn their children to swim in clay holes, that are filled in wet seasons by rain.

They are the most expert of any perhaps in America in tracking what they are in pursuit of, and they will follow their flying enemy, on a long gallop, over any kind of ground without mistaking.

Since I am on this subject, I cannot forbear taking notice of one thing related by many writers on America; which is the knowledge the savages have by the track of what kind of people they pursue; this is very true, and this sagacious particular deserved admiration, but the wonder must cease when I tell my reader, that I have found in it much of a juggle, for instead of knowing it by the foot steps (which they pretend to measure very ceremoniously with their hands) they know it by the strokes of the hatchets in the trees and branches as they go along, which no two savage nations agree in, be it in the height from the ground, or in the slope of the cut; they can also distinguish the different ways of making camps and fires; for instance; a Chactaw war camp is circular, with a fire in the center, and each man has a crutched branch at his head to hang his powder and shot upon, and to set his gun against, and the feet of all to the fire; a Cherokee war camp is a long line of fire, against which they also lay their feet; a Chactaw makes his camp in travelling in form of a sugar loaf; a Chicasaw makes it in form of our arbours; a Creek like to our sheds, or piazzas, to a timber house; in this manner every nation has some distinguishing way.

The Chicasaws are esteemed good hunters, they have extensive hunting grounds, and make excellent use of them; they extend them to the branch of the *Ohio*, called *Tanasse, Hogoheechee* or *Cherokee* river, and claim to the mouth of the *Ohio*, but this ground they frequent with great caution, only in the depth of winter when their northern enemies are close at home; they are often surprised on the rivers *Margot* and *Yasoo*, but below the *Yasoo* as far east as the eastern branches of *Tombechbé*, and as low as *Oka Tibehaw*, they hunt safely; this last they regard as their boundary with the Chactaws, but these two nations are by no means jealous of each other in this respect, and hunt in each others grounds without lett or hindrance from either side; although their country abounds in beaver, they kill none, leaving that to the white men; they think this kind of hunting beneath them, saying any body can kill beaver, but men only deer; this is exactly the reverse of a northern Indian; they hunt like all their neighbours with the skin and frontal bone of a deer's head, dried and stretched on elastic chips; the horns they scoup out very curiously, employing so much patience on this, that such a head and antlers often do not exceed ten or twelve ounces; they fix this on the left hand, and imitating the motions of the deer in fight, they decoy them within sure shot. I cannot forbear to mention a merry accident on this occasion; a Chactaw Indian, who was hunting with one of these decoys on his fist, saw a deer, and thinking to bring it to him, imitated the deer's motions of feeding and looking round in a very natural way, another savage within shot, mistaking the head for a real one, shot the ball through it, scarcely missing the fingers of the first; the affair ended in fisty cuffs, but was no farther resented.

Their habitations at home consist of three buildings, a summer house, a corn house, and a winter house, called a hot house; the two first are oblong squares, the latter is circular, they have no chimnies but let the smoke find its way out through a hole at the top in their dwelling houses, but in the hot houses, where it can; in these they make large wood fires, on the middle of the floor, which being by evening all coals, they enter in, and sleep on benches made round the inside of the building; this would stifle any one not used to it, and be it never so sharp a morning, they come out sweating and naked as soon at it is day; I believe this proceeding kills numbers of them, as in latitude 35 00, where they live, it is often very cold; they also use for an universal cure of all diseases, excessive sweating in these hot houses, and then with their pores open jump into a hole of cold water, this treatment of those that had the small pox killed numbers; these hot houses of a morning emitting smoke through every crevice, seem to a stranger to be all on fire on the inside.

Their common food is the *zea* or the Indian corn, of which they make meal, and boil it; they also parch it, and then pound it; thus taking it on their journey, they mix it with cold water, and will travel a great way without any other food; they begin to have the knowledge of keeping cattle; but at present they enjoy little or no fresh meats while at home, but in the hunting season in the woods, it is almost the only food they make use of; they have also a way of drying and pounding their corn, before it comes to maturity; this they call *Boota Copassa* (i.e.) cold flour; this, in small quantities, thrown into cold water, boils and swells as much as common meal boiled over a fire; it is hearty food, and being sweet, they are fond of it; but as the process for making it is troublesome, their laziness seldom allows them to have it; they likewise use hickory nuts in plenty, and make a milkey liquor of them, which they call milk of nuts; the process is at bottom the same as what we use to make milk of almonds; this milk they are very fond of, and eat it with sweet potatoes in it; they also make a great use of Bears fat as oil; the flesh the traders have learned them to make into bacon, exactly resembling that of a hog; but all these dishes suit but ill the palate of an European, and when they have any deer or buffalo flesh at home, it is so dried as to have no taste in it.

The knowledge they have of cattle keeping is borrowed from the traders among them, who, notwithstanding the ordinance against settling on Indian grounds, have many of them plantations, and raise cattle and hogs; one Caldwell has the greatest stock; and *Opaya Mingo Luxi* went in 1771 to complain of it, but Caldwell, knowing that no savage can withstand the words of a white man, took advantage thereof, and so intimidated the savage, by his meer presence at Pensacola, when in the superintendent's hall, in order to lodge his information, and make his complaint that *Opaya Mingo Luxi* himself said he had nothing against him; but as the very Commissary has a plantation and cattle, and keeps negroes &c. for the cultivation thereof (though he keeps his cattle under the name of *Opaya Mataha*) I think very little will be done to hinder it; and upon the whole, I think the affair of advantage to the savages, who must soon generally give into this way of life for their own preservation, or else remove further from us.

This office of Commissary seems to me the most needless expense the crown is at, as it only serves for a subject of ridicule both to the traders and savages, which last scruple not often to give the officer in this nation the (among them) scandalous epithet of old woman; and he can do but little towards preventing disorders among them, or in regulating the standard of the trade; besides, I am sure that whatever Commissiary dared to pretend to be any thing more than a cypher, would run an imminent risque of his life.

Their numbers have been very large, and they themselves have a tradition that they were a colony from another nation in the West, and that they first set themselves down near the *Ohio*, but soon removed to their present Site; the greatest number that their gunmen can now be reckoned at, does not exceed two hundred and fifty; it is really amazing, to think, that such a handfull keeps about ten thousand of the men of the other tribes from destroying them; but their ferocity and the way of making war among the savages which gives no advantage to numbers, because the war parties of a small nation are as numerous as those of a larger one) has long saved them from destruction.

Strong liquors make a sad havock among these as among all other nations of the savages in the North.

They are strong, and swift of foot, and their exercise at home is chiefly their ball play, a very laborious diversion.

They are horridly given to sodomy, committing that crime even on the dead bodies of their enemies, thereby (as they say) degrading them into women.

In their war parties, they have generally one who has done most mischief to the enemy for their leader; but he is so far from having a command, that an attempt, to do more than proposing whether such or such an undertaking would not be most adviseable, or at most persuading them to it, would at least be followed by a total desertion.

They are very ceremonious in their preparations for war, and their fondness for witchcraft makes them look for omens of futurity.

They and all other savages have the greatest share of patience imaginable; when a scalp or prisoner is in question, they will travel hundreds of miles in the desarts, with amazing precaution, enduring hunger, and often thirst, at a great rate; nay if their provisions fail before they strike the blow, they have been known to return to hunt for more in some safe place, and without going home, to make a second or third attempt.

They make war by stratagem, surprise or ambush, despising us as fools for exposing ourselves to be shot at like marks. A man's valour with them consists in their cunning, and he is deemed the greatest hero who employs most art in surprising his enemy; they never strike a blow unless they think themselves sure of a retreat, and the loss of many men is an infamous crime laid to the charge of the party.

They bury their dead almost the moment the breath is out of the body, in the very spot under the couch on which the deceased died, and the nearest relations mourn over it with woeful lamentations; the women are very vociferous in it, but the men do it in silence, taking great care not to be seen any more than heard at this business; the mourning continues about a year, which they know by counting the moons, they are every morning and evening, and at first throughout the day at different times, employed in the exercise of this last duty.

A people who by many peculiar customs, are very different from the other red men on the continent, will next amuse us: They are the Chactaws, more commonly known by the name of the Flatheads. These people are the only nation from whom I could learn any idea of a traditional account of a first origin; and that is their coming out of a hole in the ground, which they show between their nation and the Chicasaws; they tell us also that their neighbours were surprised at seeing a people rise at once out of the earth; dark as this account of a first existence of these people is to us, we discover in it a higher idea of an origin than among their neighbours, who never pretend to tell from whence they came, and have only loose ideas of a migration from the north west; which prevail among the Chicasaws and all other southern nations.

The Chactaws may more properly be called a nation of farmers than any savages I have met with; they are the most considerable people in Florida, and their situation may be known by the annexed plan of their country: Their hunting grounds are in proportion less considerable than any of their neighbours; but as they are very little jealous of their territories, nay with ease part with them, the Chicasaws and they never interrupt each other in their hunting; as I mentioned before.

They are in their warlike temper far from being such cowards as people in general will pretend, but it is true they are not so fond of wandering abroad to do mischief as the other savages are; few of such expeditions are undertaken by them, and they give for a reason, that in going abroad they may chance to be obliged to content themselves with a woman's or child's scalp, but in staying at home and waiting attack of the enemy, they by pursuing them, are sure to take men, which is a greater mark of valour: be this as it will, it is certain they are carefully, cunningly, and gravely watchfull at home, and on several occasions they have, after many insults, boldly offered to meet their enemies in equal numbers on a plain, which has always been by the other savages treated with scorn, as cowardice; however when it has happened by chance that they meet so, we have seen them brave and victorious. Even in the very town of Mobile, an action of this kind happened deserving a record, when they drove their enemies (the Creeks) through the river, and but for their inability to swim, they would have totally destroyed them; the Captain *Hooma* or red Captain fighting with forty Men against three hundred Creeks, and with his own hand destroying thirteen of their Chiefs, even when fighting on his knees, and when he fell, bravely telling who he was, and his being flead alive for his heroism, is so fresh in every one's memory (being not above six Years ago) that many living evidences can testify it; I thought the action worthy of this attempt, to save it from oblivion. They have deserted many of their eastern frontier towns since their present war with the Creeks, but during my stay in their nation, I saw four or five instances of their not suffering their enemy to escape unpunished, when he dared to commit depredations, and they valued themselves on the event of the present war, when in 1771, news coming among the Traders, that the Creeks computed their loss at near three hundred persons, and they having guessed the number of their's, lost much the same; they said, we have lost many women and children and even of them some Scalps have been retaken, but we like men, have killed men only, and got all the marks thereof; this war began in August 1765; the readers may judge at the greatness of their exploits, when I assure them, that that number was the total loss during all that time.

These savages were the staunch and firm friends of the French while they continued on the continent, until some English traders found means to draw the east party, and the district of *Coosa* (which together are called *Oypat-oocooloo*, or the small nation) into a civil war with the western divisions called *Oocooloo-Falaya, Oocooloo-Hanalé*, and *Chickasawhays*, which after many conflicts and the destruction of east *Congeeto*, ended with the peace in 1763. I believe they are a nation whose word may be depended on when they give into the interest of any person, and that their faith is to be better relied on then that of the Chicasaws or Creeks, which two last are really versed in all the gallic tricks of deceit.

At the congress of 1771, there were two thousand three hundred of this nation nearly all men, at Mobile on the superintendants books; in the nation I found at above seventy of their villages, about two thousand men, and in the woods and hunting grounds, I was at and heard of as many camps as could make no less than six hundred men more. The French used to keep them very poor, but the yearly small gifts they were accustomed to, were more proper to leave an idea of gratitude on the mind of a savage, than our septennial great ones.

Monsieur de Kerlerec and others made a fine juggle and kind of monopoly of this trade, which was very ill brooked by the French; but although I make no doubt of the Gentlemen's having gone beyond the orders of their court, and notwithstanding I am a bitter enemy to unreasonable regraters, yet I am certain that a monopoly of this trade under proper restrictions would prove an advantage and security to the colonies, since now the villainous over reaching, chicanery, and mutual calumniations of the abandoned wretches, who reside among the savages, joined to their worse than brutish or savage way of life, tend to the rendering the nation to which they belong infamous, despicable, and scandalous among the savages, as well as to turn the hopes of advantage into a real disadvantage; and I dare venture to say, that unless such restrictions take place, the different savages will always find reason to complain against the colonies, and join in cabals against the poor settlers in the remote counties, and at last oblige the colonies to take the disagreeable step of realizing our mock authority, by extirpating all savages that dare to remain on the East of the Mississippi.

But to return to the war-like inclinations of the Chactaws, and not to involve myself in politicks; I take them to be a brave people, who can upon occasion defend themselves very coolly, for during my stay in the nation, a woman of that tribe made a bargain with me to give her ammunition for some provisions I bought of her; and when I expressed my surprise thereat, she informed me that she kept a gun to defend herself as well as her husband did; and I have several times seen armed women in motion with the parties going in pursuit of the invading enemy, who having completed their intended murder, were flying off.

They never exercised so much cruelty upon their captive enemies as the other savages; they almost always brought them home to show them, and then dispatched them with a bullet or hatchet; after which, the body being cut into many parts, and all the hairy pieces of skin converted into scalps, the remainder is buried and the above trophies carried home, where the women dance with them till tired; then they are exposed on the tops of the hot houses till they are annihilated. The same treatment is exercised on those who are killed near the nation, but he that falls in battle at a distance is barely scalped.

Their addictedness to pretended witchcraft leads them into a very superstitious behaviour when on an expedition which is remarkable, they carry with them a certain thing which they look on as the genius of the party; it is most commonly the stuffed skin of an owl of a large kind; they are very careful of him, keep a guard over him, and offer him a part of their meat; should he fall, or any other ways be disordered in position, the expedition is frustrated; they always set him with his head towards the place of destination, and if he should prove to be turned directly contrary, they consider this as portending some very bad omen, and an absolute order to return; should therefore any one's heart fail him, he needs only watch his opportunity to do this to save his character of a brave or true man. There is also a species of *Motacilla* (which I often endeavoured to catch, in vain) whose chirping near the camp, will occasion their immediate return.

Their war camps are already mentioned.

They are given to pilfering, but not so much as the Chicasaws.

They are the swiftest of foot of any savages in America, and very expert in tracking a flying enemy, who very seldom escapes.

Their leader can not pretend to *command* on an expedition, the most he can do, is to endeavour to *persuade*, or at the extent, he can only pretend to a greater experience in order to enforce his counsel; should he pretend to order, desertion would at least be his punishment, if not death.

Their exercises agree pretty much with what I have seen among other nations: from their infancy they learn the use of bows and arrows; they are never beaten or otherways rudely chastised, and very seldom chid; this education renders them very willful and wayward, yet I think it preferable to the cruel and barbarous treatment indiscriminately used by some European parents, who might with slight punishments by the excellency of wholesome christian Admonitions, work in a very different manner on the tender inclinations of pliable infancy.

The young savages also use a very strait cane eight or nine feet long, cleared of its inward divisions of the joints; in this they put a small arrow, whose one end is covered one third of the whole length will cotton or something similar to it; this they hold nearest their mouth and blow it so expertly as seldom to miss a mark fifteen or twenty yards off and that so violently as to kill squirrils and birds therewith; with this instrument they often plague dogs and other animals according to the innate disposition to cruelty of all savages, being encouraged to take a delight in torturing any poor animal that has the misfortune to fall into their hands; thinking best of him, who can longest keep the victim in pain, and invent the greatest variety of torture. When growing up, they use wrestling, running, heaving and lifting great weights, the playing with the ball two different ways, and their favourite game of *chunké*, all very violent exercises.

The excess in spirituous liquors to which they use themselves, is really incredible.

Their meetings about serious matters are at night. Their belief in charms and exorcisms is firm and out of reason, and he that should dare openly to boast of this gift is sure to lose his life on the first misfortune in the town where he resides; but if it is only pretended to extend to the cure of wounds and diseases it is overlooked; when they prepare for war, and when they return they use exorcisms, they call them all physic though only bare words or actions; and if they prove unsuccessful, they say the physic was not strong enough; it is no small diversion to see a Chactaw during this preparation act all his strange gestures, and the day before his departure painted scarlet and black almost naked and with swan wings to his arms run like a bacchant up and down through the place of his abode; not drunk neither, as rum is by them avoided like poison during this preparation.

While I was in this nation, I had the misfortune to be afflicted with a violent fever which ended in a flux; my own skill being baffled, I applied to my guide, who had the reputation of being a knowing Physician well acquainted with the simples used among them. I submitted to his prescription; he got some herbs and roots, and made a decoction of them; I drank it; while the effect was expected, he alternately burnt some of the simples and sat down by me blowing upon me to drive away the disorder; I found no benefit by it; and on my refusing an other trial he said I was a fool, the next time the physic would be stronger, but he was not affronted.

The French have made great attempts to render his nation Christians; at *Chicasawhay* there resided a missionary and a chapel was built, but they were absolutely unsuccessful; the savages always derided the Jesuit, called him a woman, and would frequently desire him to take away his physic, thereby meaning he would undo his ceremony of baptism; and when the English arrived there they would go up to the altar and imitate all the jesuitical farce, telling them that they were not such fools as to hear him; the chapel was destroyed before I came there in 1771, but the cross (being of lightwood*) stood yet.

Their play at ball is either with a small ball of deer skin or a large one of woollen rags; the first is thrown with battledores, the second with the hand only; this is a trial of skill between village and village; after having appointed the day and field for meeting, they assemble at the time and place, fix two poles across each other at

*The heart of yellow pine.

about an hundred and fifty feet apart, they then attempt to throw the ball through the lower part of them, and the opposite party trying to prevent it, throw it back among themselves, which the first again try to prevent; thus they attempt to beat it about from one to the other with amazing violence, and not seldom broken limbs or dislocated joints are the consequence; their being almost naked, painted and ornamented with feathers, has a good effect on the eye of the by stander during this violent diverision; a number is agreed on for the score, and the party who first gets this number wins.

The women play among themselves (after the men have done) disputing with as much eagerness as the men; the stakes or betts are generally high. There is no difference in the other game with the large ball, only the men and women play promiscuously, and they use no battledores.

Their favourite game of *chunké* is a plain proof of the evil consequences of a violent passion for gaming upon all kinds, classes and orders of men; at this they play from morning till night, with an unwearied application, and they bet high; here you may see a savage come and bring all his skins, stake them and lose them; next his pipe, his beads, trinkets and ornaments; at last his blanket, and other garment, and even all their arms, and after all it is not uncommon for them to go home, borrow a gun and shoot themselves; an instance of this happened in 1771 at *East Yasoo* a short time before my arrival. Suicide has also been practised here on other occasions, and they regard the act as a crime, and bury the body as unworthy of their ordinary funeral rites.

The manner of playing this game is thus: They make an alley of about two hundred feet in length, where a very smooth clay ground is laid, which when dry is very hard; they play two together having each a streight pole of about fifteen feet long; one holds a stone, which is in shape of a truck, which he throws before him over this alley, and the instant of its departure, they set off and run; in running they cast their poles after the stone, he that did not throw it endeavours to hit it, the other strives to strike the pole of his antagonist in its flight so as to prevent its hitting the stone; if the first should strike the stone he counts one for it, and if the other by the dexterity of his cast should prevent the pole of his opponent hitting the stone, he counts one, but should both miss their aim the throw is renewed; and in case a score is won the winner casts the stone and eleven is

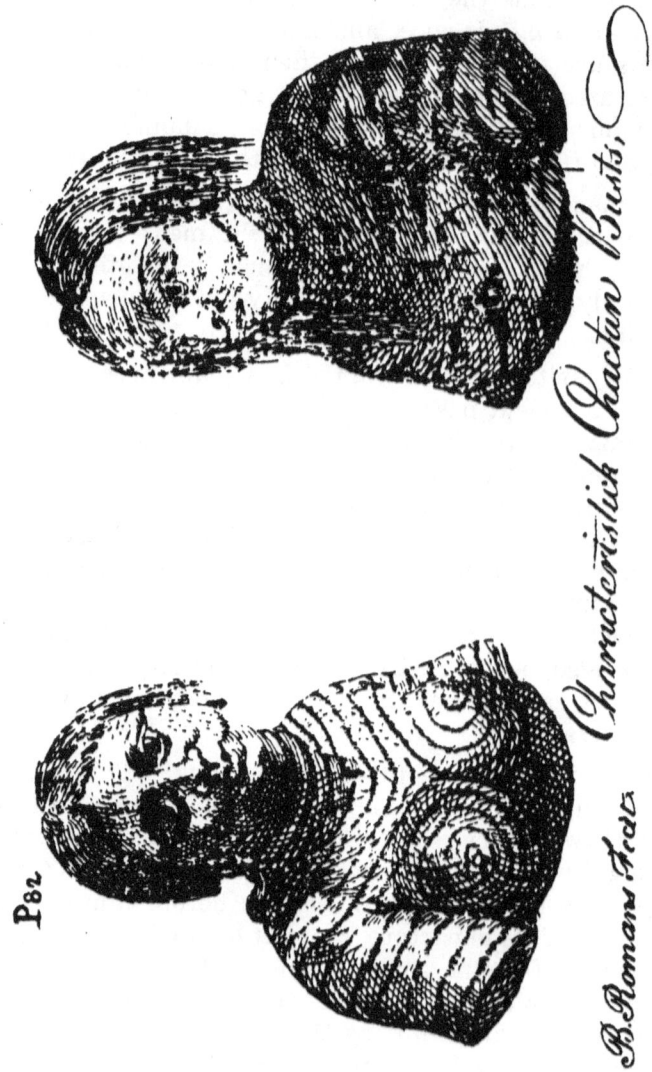

B. Romans Fecit. Characteristich Chactaw Busts.

up; they hurl this stone and pole with wonderful dexterity and violence, and fatigue themselves much at it.

The women also have a game where they take a small stick, or something else off the ground after having thrown up a small ball which they are to catch again, having picked up the other; they are fond of it, but ashamed to be seen at it. I believe it is this propensity to gaming which has given these savages an idea of a *meum* and *tuum* above all other nations of America.

They are extravagant in their debauches; when met for a drinking match some women attend them, when these find the men beginning to be heated with liquor they will take away all the weapons found near them and return with a callebash under their wrappers, then mixing with them, the men offer them their bottles, they take a draught and when not observed they empty it into the callebash, which when full they empty into bottles brought for that purpose, and thus they will accumulate two or three bottles full, and with the help of a little water, still make them more; after a while rum fails among the men, and the women acquaint them, that they have got some; they are told to fetch it; they refuse, saying it cost them much and they cannot give it for nothing; a bargain ensues, they receive the consideration first, and then bring it; in this way of trade they will often get all the effects the men can command for such a delicate nectar.

I have a great opinion of a Chactaw's faithfully performing his promises. I have seen several little instances thereof; they detest a liar, and shew gratitude to a man that keeps his word; my guide whose name was *Pooskoos Mingo* gave me an instance of this; when I left him he said I had satisfied him for every thing like a true man, but if I would give him a speaking paper to the great white man at Mobile (meaning John Stuart Esq) then he would still better know it; I gave him a note recommending him to that gentleman, and because he had been of extraordinary service on the journey, begged he would allow him something more than common; it had the desired effect, he got a good many things extraordinary; when I was afterwards missing, and it was thought the Creeks had destroyed us in coming from the Chicasaw nation, this savage armed to avenge my death, and was actually taking the war physick as they term it, when news was brought to the nation of my arrival at Mobile.

They are well made both men and women; the women have agreeable features and countenances, but their nasty way of life in general disfigures them; those that are cleanly are realy attractive; the women disfigure the heads of their male children by means of bags of sand, flattening them into different shapes, thinking it adds to their beauty; both men and women wear long hair, except some young fellows who begin to imitate the Chicasaw fashion, and both sexes mark their faces and bodies, particularly the women with indelible blue figures of fancy, among which is a great deal of voluted work of vast variety.

Before the English traders came among them, there were scarcely any half breed, but now they abound among the younger sort.

Both sexes are wanton to the highest degree, and a certain fashionable disorder is very common among them. Sodomy is also practised but not to the same excess as among the Creeks and Chicasaws, and the *Ginaedi* among the Chactaws are obliged to dress themselves in woman's attire, and are highly despised especially by the women.

Their buildings are exactly similar to those of the Chicasaws.

Their way of life in general may be called industrious, they will do what no uncompelled savage will do, that is work in the field to raise grain; and one may among them hire not only a guide, or a man to build a house, or make a fence, but even to hoe his grounds; nay they will for payment be your menial servants to the meanest offices; no other unsubdued savage will do any more of all these than be your guide; they are very ingenious in making tools, utensils and furniture; I have seen a narrow tooth comb made by one of these savages with a knife only out of a root of the *Diospyros* that was as well finished as I ever saw one with all the necessary tools; this shews their patience.

The Chactaws are very hospitable at their hunting camps, and there only they will entertain a stranger at free cost.

Here I must relate a particular custom of these people: When a deer or bear is killed by them, they divide the liver into as many pieces as there are fires, and send a boy to each with a piece, that the men belonging to each fire may burn it, but the women's fires are excluded from this ceremony, and if each party kills one or more animals, the livers of them are all treated in the same manner.

Horses of a good kind are in such plenty as to be sold for a kegg of four gallons half water rum; they would be excellent were it not that they back them before the animals attain two years of age.

They cultivate for bread all the species and varieties of the *Zea*, likewise two varieties of that species of *Panicum* vulgarly called guinea corn; a greater number of different *Phaseolus* and *Dolichos* than any I have seen elsewhere; the esculent *Convolvulus* (vulgo) sweet potatoes, and the *Helianthus Giganteus;* with the seed of the last made into flour and mixed with flour of the *Zea* they make a very palatable bread; they have carried the spirit of husbandry so far as to cultivate leeks, garlic, cabbage and some other garden plants, of which they make no use, in order to make profit of them to the traders; they also used to carry poultry to market at Mobile, although it lays at the distance of an hundred and twenty miles from the nearest town; dunghill fowls and a very few ducks, with some hogs, are the only esculent animals raised in the nation.

They make many kinds of bread of the above grains with the help of water, eggs, or hickory milk; they boil the esculent convolvulus and eat it with the hickory milk; they boil green ears of corn, they boil corn and beans together, and make many other preparations of their vegetables, but fresh meat they have only at the hunting season, and then they never fail to eat while it lasts; of their fowls and hogs they seldom eat any as they keep them for profit.

In failure of their crops, they make bread of the different kinds of *Fagus*, of the *Diospyros*, of a species of *Convolvulus* with a tuberous root found in the low cane grounds, of the root of a species of *Smilax*, of live oak acorns, and of the young shoots of the *Canna;* in summer many wild plants chiefly of the *Drupi* and *Bacciferous* kind supply them.

They raise some tobacco, and even sell some to the traders, but when they use it for smoking they mix it with the leaves of two speices of the *Cariaria* or of the *Liquidambar Styraciflua* dried and rubbed to pieces.

They prepare a kind of cloth out of the bark of a species of *Morus*, and with its root dye it yellow; I have all the reason in the world to believe, that this cloth might be manufactured into paper.

Buffaloe's wool also furnishes them a material for a useful manufacture.

They likewise make blankets and other coverings out of the feathers of the breasts of wild turkies by a process similar to that of our wig makers, when they knit hair together for the purpose of making wigs.

They have a root by means of which they dye most substances of a bright lasting scarlet, but hitherto I have not been able to discover what it was.

Many among them are well acquainted with plants of every kind, and apply them judicially both externally and internally; to others again they attribute supernatural virtues; for instance, there is one which they make use of to procure rain; for this purpose they have a number of people in their nation called rainmakers; these assemble in a deserted field, and they boil this plant in a large pot, dancing and singing round it with numberless aukward gestures; then if it should happen to rain soon after, the jugglers boast the virtue of the plant; but should no rain follow, they say the physick was not strong enough; they take care however not to employ this rain compelling herb unless a cloudy day forebodes rain. The plant is very singular, and I believe a nondescript; I saw two species of it, but could not ascertain the genius; the savages call it *Esta Hoola* or the most beloved.

The most remarkable thing of these savages is their inability to swim, occasioned by their being remote from large waters; this art the people of *Chicasahay* and *Yoani* who live on the banks of the *Pasca Oocooloo* enjoy alone, and incredible as it may appear, even most of their horses partake of this inability, as many people and among others the Commissary for the nation have informed me from their own exeperience.

They help their wives in the labour of the fields and many other works; near one half of the men have never killed a deer or turkey during their lives. Game is so scarce, that during my circuit through the nation we never saw any, and we had but two or three opportunities of eating venison in as many months; they take wives without much ceremony, and live together during pleasure, and after separation which is not very frequent, they often leave the second to retake the first wife.

Fornication is among them thought to be a natural accident, therefore a girl is not the worse looked on for ten or a dozen slips; but although they are not over jealous of their wives, they punish adultery in the woman, unless she happens to belong to a stronger or more noted and numerous family than the husband; in which case he scarce ventures even to put her away; but if she is doomed to suffer, her punishment is to be at a publick place (for that purpose set apart at every town) carnally known by all who choose to be present, young and old; thus the poor wretch after defending herself and struggling hard with the first three or four, at last suffers motionless the brutality of perhaps an hundred or an hundred and fifty of these barbarians; the same treatment is undergone by a girl or woman who belonging to another town or quarter of the nation, comes to a place where she is a stranger and cannot give a very good account of herself and business, or the reason of her coming there; this they call running through the meadow, and if a white man happens to be in the town, they send him an offer or invitation to take the first heat; they plead in excuse for so barbarous a custom that the only way to disgust lewd women is to give them at once what they so constantly and eagerly pursue.

The education of their children I have already mentioned.

The women suffer no more by child birth than any other savage women; they retire into a place of solitude at the time, and after delivery return to their daily labour; however, while I staid at *Oka Altakkala* in this nation one died in labour within about eighty yards of the house I resided in.

There are no laws or regulations observable among these people, except the *Lex Talionis*, and although they have a strict notion of distinction in property, and even divide their lands, we never hear them quarrel about boundaries; the above law is so strictly followed, that I am furnished with the following anecdote: It happened that a young Chactaw having done something deserving reproof, he was therefore chid by this mother, this he took so ill as in the fury of his shame to resolve his own death, which he effected with a gun; his sister as his nearest relation thought herself bound to avenge his death, and knowing the circumstances told her mother she had caused her brother's death and must pay for his life; the old woman resigned herself to her fate, and died by the hands of her daughter, who shot her with a gun which she had provided for the catastrophe.

In sickness the juggling Quacks are consulted, and as they are naturally good connoisseurs in simples, and judge pretty well of the nature of diseases, they often succeed; but if a disorder is obstinate or incurable, the relations of the patient assemble in his house, bewail his misfortune, cry bitterly, take their leave of him, and he tells them how tired he is of life, that his misfortunes are unsufferable, and that it is good he should die; upon this an universal howl is raised, the nearest male relation jumps on him, and violently in a moment breaks the neck of the patient, and then they rejoice that his misery is over, but lamentations for his departure soon succeed.

The following treatment of the dead is very strange, yet we find *Apollonius Rhodius* mention a similar custom of the inhabitants of *Colchis* near *Pontus;* we find *Ives* in his voyage relating the like of the remainder of the ancient Persians, and we find again in *Hawkesworth's* voyages the people of *Otaheite,* performing their obsequies in a manner little or nothing different from the Chactaws; but it would be an hard matter to assign a reason for it among the latter; that it is not with any solicitude about the disposition of the body in regard to a future state is plain; all the reason I could hear of them was, that they would not so soon forget their deceased friends, and might be the oftener stimulated to weep over their remains. As soon as the deceased is departed, a stage is erected (as in the annexed plate is represented, and the corpse is laid on it and covered with a bear skin; if he be a man of note, it is decorated, and the poles painted red with vermillion and bears oil; if a child, it is put upon stakes set across; at this stage the relations come and weep, asking many questions of the corpse, such as, why he left them? did not his wife serve him well? was he not contented with his children? had he not corn enough? did not his land produce sufficient of every thing? was he afraid of his enemies? &c. and this accompanied by loud howlings; the women will be there constantly, and sometimes with the corrupted air and heat of the sun faint so as to oblige the by standers to carry them home; the men will also come and mourn in the same manner, but in the night or at other unseasonable times, when they are least likely to be discovered.

The stage is fenced round with poles, it remains thus a certain time but not a fixed space, this is sometimes extended to three or four months, but seldom more than half that time. A certain set of venerable old Gentlemen who wear very long nails as a distinguishing badge on the thumb, fore and middle finger of each hand, constantly travel through the nation (when I was there, I was told there were but five of this respectable order) that one of them may acquaint those concerned, of the expiration of this period, which is according to their own fancy; the day being come, the friends and relations assemble near the stage, a fire is made, and the respectable operator, after the body is taken down, with his nails tears the remaining flesh off the bones, and throws it with the intrails into the fire, where it is consumed; then he scrapes the bones and burns the scrapings likewise; the head being painted red with vermillion is with the rest of the bones put into neatly made chest (which for a Chief is also made red) and deposited in the loft of a hut built for that purpose, and called bone house; each town has one of these; after remaining here one year or thereabouts, if he be a man of any note, they take the chest down, and in an assembly of relations and friends they weep once more over him, refresh the colour of the head, paint the box red, and then deposit him to lasting oblivion.

An enemy and one who commits suicide is buried under the earth as one to be directly forgotten and unworthy the above ceremonial obsequies and mourning.

A mixture of the remains of the *Cawittas, Talepoosas, Coosas, Apalachias, Conshacs* or *Coosades, Oakmulgis, Oconis, Okchoys, Alibamons, Natchez, Weetumkus, Pakanas, Taënsas, Chacsihoomas, Abékas* and some other tribes whose names I do not recollect, will be the next subject of our attention; they call themselves *Muscokees* and are at present known to us by the general name of *Creeks,* and divided into upper and lower Creeks; also those they call *allies* and are a colony from the others living far south in East Florida.

They inhabit a noble and fruitful country, where a civilized people in future will enjoy all the earthly sweets they can wish for, and where the inhabitants will always be placed commodiously for navigation, so as with little trouble to bring all the valuable produce of a rich soil situate in a temperate air of the middle latitudes to a market; in a word, I foresee this will become the seat of trade and its attendant riches in North America.

They are the next most numerous nation after the *Chactaws;* but because I have not been so universally through this nation as through the others, I cannot so nearly calculate their numbers, but to all appearance, three thousand five hundred gun men is all the extent; this confederacy of remnants is a race of very cunning fellows, and with regard to us, the most to be dreaded of any nation on the continent, as well for their indefatigable thirst for blood (which makes them travel incredibly for a scalp or prisoner) as for their being truly politicians bred, and so very jealous of their lands, that they will not only not part with any, but endeavour constantly to enlarge their territories by conquest and claiming large tracts from the Cherokees and Chactaws.

As an instance of their politicks, I beg leave to relate the following fact: When in 1764 and 1765, Messrs. Rea and Galphin of Georgia, had the contract of providing Pensacola with beef they were of necessity obliged to have the cattle drove through this nation, who never suffered any but *oxen* to pass, the tendency of this prohibition must be obvious to every reader.

They are all remarkably well shaped, they live in a level country full of rivers, are expert swimmers, and in general a very hardy race; what deserves notice here is that their *Thorex* is very shallow, so that a savage of this race may appear almost a giant by the breadth of his shoulders and yet not measure so much in circumference as an ordinary European; but whether this is the effect of art or nature, I cannot pretend to decide; their women are handsome and many of them very cleanly, they are very hospitable and never fail of making a stranger heartily welcome, offering him the pipe as soon as he arrives, while the good women are employed to prepare a dish of venison and homany*, with some bread made of maize and flour, and being wraped in maize leaves, baked under the ashes; when it is served up they accompany it with bears fat purified to a perfect chrystalline oyl, and a bottle of honey with which last article the country abounds, and it is of so good a quality, as in my opinion to exceed that of *Calabria* and *Minorca*.

In the lower nation and the allied tribes, there are many who keep rice by them and have plenty of beef; of all which articles they are profusely liberal, and I believe had they only a single potatoe, they would share it with a stranger. In the fruit season they never fail to accompany these regales with melons, peaches, plumbs, grapes, or some other wild fruit.

After the traveller is made welcome by his host, the latter introduces him into the assembly, which is kept every evening at a place called the square, of which we find one in each town. At this place he is entertained with tobacco and caffine drink; this is also the common resort of their old men and warriors to deliberate on matters of peace and war, to judge what steps are to be taken for the welfare of the nation, and to decide the fate of their neighbours; so that this square doth not ill answer to the description we have of the Roman *Forum* or Athenian *Areopage;* the evening ends in a dance, which is the common practice every night.

To relate any thing concerning the wars of this nation, would be no more than repeating what has been said of the Chicasaws.

Their way of life is in general very abundant; they have much more of venison, Bear, turkies; and small game in their country than their neighbours have, and they raise abundance of small cattle,

*Maize coarsly pounded, sifted and boiled in water.

B. Roman fecit

Characteristick head of a Creek War Chief.

hogs, turkeys, ducks and dunghill fowls (all which are very good in their kind) and of these they spare not; the labour of the field is all done by the women; no savages are more proud of being counted hunters, fishermen, and warriors: were they to cultivate their plentiful country, they might raise amazing quanties of grain and pulse, as it is they have enough for their home consumption, they buy a good deal of rice, and they are the only savages that ever I saw that could bear to have some rum in store; yet they drink to excess as well as others; there are few towns in this nation where there is not some savage residing, who either trades of his own stock, or is employed as a factor. They have more variety in their diet than other savages: They make pancakes; they dry the tongues of their venison; they make a caustick salt out of a kind of moss found at the bottom of creeks and rivers, which although a vegetable salt, does not deliquiate on exposing to the air; this they dissolve in water and pound their dried venison till it looks like oakum and then eat it dipped in the above sauce; they eat much roasted and boiled venison, a great deal of milk and eggs; they dry peaches and persimmons, chestnuts and the fruit of the *chamaerops*, they also prepare a cake of the pulp of the species of the *passi flora*, vulgarly called may apple; some kinds of acorns they also prepare into good bread; the common esculent *convolvulus* and the sort found in the low woods, both called potatoes, are eat in abundance among them; they have plenty of the various species of *Zea* or maize, of the *Phaseolis* and *Dolichos*, and of different kinds of *Panicum;* bears oyl, honey and hickory milk are the boast of the country; they have also many kinds of salt and fresh water turtle, and their eggs and plenty of fish; we likewise find among them salted meats, corned venison in particular, which is very fine; they cultivate abundance of melons; in a word, they have naturally the greatest plenty imaginable; were they to cultivate the earth they would have too much; vast numbers of horses are bred here, but of an indifferent kind; and these savages are the greatest horse stealers yet known; it is impossible to be sure of a horse wherever these fellows come.

The caffine is by them used as a drink, they barbecue or toast the leaves and make a strong decoction of them; the men only are permitted to drink this liquor to which they attribute many virtues, and it is made so strong as to be black and raise a froth; when they drink it at their assemblies in the square, they call it black drink.

Every afternoon a young savage warns the village to dance; as soon at it is dusky they make a fire of dry pitch pine, and round this they dance in a circle with many strange gestures, postures, and cries; the women sing regularly and some very prettily to the musick of a kind of drum. I have heard them sing, and seen them dance to no more than the words *Yahoodela, Yahoyahena* for above two hours.

On this occasion I must not forget to mention an instance of female fondness for dress, which I saw at one of these dancing assemblies: I observed the women dressed their legs in a kind of leather stockings, hung full of the hoofs of the roe deer in form of bells, in so much as to make a sound exactly like that of the Castagnettes; I was very desirous of examining these stockings and had an opportunity of satisfying my curiosity on those of my landlady at her return home. I counted in one of her stockings four hundred and ninety three of these claws; there were nine of the women at the dance with this kind of ornament, so that allowing each of them to have had the same number of hoofs, and eight hoofs to a deer, there must have been killed eleven hundred and ten deer to furnish this small assembly of ladies with their ornaments, besides which, earrings, bracelets, &c. are by no means forgot; an instance of luxury in dress scarcely to be paralleled by our European ladies.

The men are also very fond of dress; my guide across the Peninsula, employed above two hours at his toilet, at Mr. Moultrie's house, four miles from St. Augustine, before he would venture to shew himself in town.

Their principal exercises at home are ball playing in the manner afore related, and the just mentioned dances; the women are employed, besides the cultivation of the earth, in dressing the victuals, preparing, scraping, braining, rubbing and smoking the Roe skins, making macksens of them, spinning buffaloe wool, making salt, preparing caffine drink, drying the *chamaerops* and *passiflora,* making cold flour for travelling, gathering nuts and making their milk; likewise in making baskets, brooms, pots, bowls and other earthen and wooden vessels.

They live nearly in the same kind of habitations as the two other nations already mentioned, except that their hot houses are not circular but oblong squares; they learn their boys from their youth to endure all manner of hardships particularly swimming in the coldest weather; they make them frequently undergo scratching from head to foot through the skin with broken glass or gar fish teeth, so as to make them all in a gore of blood, and then wash them with cold water; this is with them the *Arcanum* against all diseases, but when they design it as a punishment to the boys, they dry scratch them (i.e.) they apply no water after the operation, which renders it very painful; they endeavour as much as possible to teach them all manner of cruelty, by making them exercise it on the poor brute creation, in order to be the better versed in it when they want to exercise it on their own species, and others of the human genus, when they unhappily become their enemies.

As hospitable as this nation is to friends, as irreconcileably inhuman are they to their enemies; there is hardly an instance of one miserable prisoner's ever having escaped their barbarity; the torments they put the wretched victims to, are too horrid to relate, and the account thereof can only serve to make human nature shudder.

No nation has so contemptible an opinion of us as these. They practice unnatural commerce with their own sex to as high a degree as the Chicasaws; they, like all other savages, are very fond of dogs, in so much as never to kill one out of a litter, and it is not uncommon in the nation to see a dog very lean, and so sensible of his misfortune as to seek a wall or post for his support before he ventures to bark.

With regard to the women, their girls are the most arrant prudes and coquets in the world, though they will never scruple to sell the use of their bodies when they can do it in private; a person who wishes to be accommodated here can generally be supplied for payment, and the savages think a young woman nothing the worse for making use of her body, as they term it; but it is a great falshood which has been related of these savages, that they exhort their young women to cohabitation with white men.

Polygamy is here allowed, though not generally made use of; they marry without much ceremony, seldom any more than to make some presents to the parents, and to have a feast or hearty regale at the hut of the wife's Father; when once married the women are bound to the strictest observation of obedience and conjugal fidelity, saying that she that has once sold herself, can not any more dispose of any thing whatever; and of their wives they are the most unreasonably jealous of any nation under the sun. Adultery is punished by severe flagellations and loss of hair, nose, and ears, in both parties, if they are taken; sometimes they spare the nose of the man and I have known some instances of white men having this misfortune and being obliged to apply to the Commissary, or the nearest Governor for a certificate to secure them from the imputation of the pillory.

Physick, or the knowledge of it, is another thing in which they pride themselves not a little, but they apply that name to all kinds of exorcisms, juggling and legerdemain tricks, as well as to the knowledge of diseases and the simples proper to cure them.

Once a year, about July, these people put out all the fires throughout the nation; they fast the two next two days; then the fire is lighted again according to their old fashion (i.e.) by drilling with a hard piece of wood on a soft one till it catches which soon happens; thus all the fires are again lighted and universal feasting ensues.

The women are just as easily delivered as those of the other savages, and immediately after birth the infant is plunged into cold water.

They revere old age to excess; in extreme sickness they will out of compassion break the neck of the decrepid or lingering patient.

The dead are buried in a sitting posture, and they are furnished with a musket, powder and ball, a hatchet, pipe, some tobacco, a club, a bow and arrows, a looking glass, some vermillion and other trinkets, in order to come well provided in the world of spirits.

The *Arkansas, Kanzas, Kappas,* or *Kwappas* are the only nation of any note after the three above named, with whom West Florida has any connection, notwithstanding they live on the west side of the river; they are supposed to be about four hundred gun men.

This nation is famous for the death of *Ferdinand Soto* in the year 1543, after his immense journey from the Bay of *Tampe* through *Apalachia, Pensacola, Mobile,* and the Chicasaw nation, &c. to the Mississippi, by his historian called *Rio Grande;* also the arrival of Mr. *de la Salle,* who in 1682, coming down this river and having taken formal possession of this country, went further down to its mouth, observed it in Latitude 29, went up again as far as *Illinois,* thence to *Canada* and France, from whence he returned with a small squadron in 1684 to find the mouth of the river once more, but through obstinacy passed it so far as to get into the Bay of *St. Bernard* and river *Bravo,* where losing his vessels he attempted to find the *Arkanzas* a second time, but after a long and tedious journey of almost three years he shared if possible a worse fate than *Soto,* being murdered by his own people near the nation of the *Caeni* on the 29th of May 1687; this we gather from the writings of Mr. *Joutel* who was present at all the scenes I have related, concerning this affair, and who pursued the journey to the *Arkanzas* and *Canada;* this nation were till lately the declared enemies of the Chicasaws; their country is said to be exceeding fine, but as I have not seen it I can not judge of it, I never saw above six of this race, therefore I only treat of them on account of their connection with the country I am describing.

In general their manners are very like those of all other savages, but among them I find a custom not known by the other southern nations which is that of a feast of dogs flesh at the declaration of war, which is a common practice among all the northern tribes; we are also told that an *Arkanza* is counted a warrior in his nation for having killed his enemies dog.

As many various accounts have been given of the manner of a declaration of war by different savages, I will here acquaint my reader of my having more or less personal knowledge or acquaintance with every tribe between the river St. Lawrence and the bottom of the gulph of Darien, and I think I can give the following as an account that will stand the universal test in all that extent.

When war has been resolved on by the leading men of the nation, a feast is prepared and all the chief warriors are invited; after a great deal of swilling, the cause of the assembly is eloquently made known by some Orator, and the intended expedition submitted to the consideration of the assembly; as soon as it is approved of, they paint themselves red or black, assume a very gastly appearance and behaviour, adorning themselves many uncouth ways; they dance their war dances under a continual roar of the death whoop, every one relates the great actions of his ancestors, and sings the praises of his own, as well past as future; the insults of the enemy are painted in their blackest colours, and a great deal of bloody advice to their young men with an universal cry of revenge! revenge! ends the farce, and then every one plans expeditions as he sees proper and convenient to execute.

The practice of scalping their slain enemy I believe is universal, but those in the neighbourhood of the Dutch colonies have laid it by for another, that of cutting off the hands, which was in consequence of a reward the company gave them for the hands of such runaway blacks as joined the rebel Negroes, because there were instances of persons scalped who have recovered, but after the loss of hands they cannot any more use arms.

Juggling and pretence to witchcraft, as also consulting of omens, is common to the *Arkanzas* as well as others, and passes by the name of physick.

They have so far a veneration for the alligator as not to destroy him, nor have I seen a savage who would willingly kill a snake.

The pipe is used here as with others, Tobacco in some shape or other seems to be the American symbol of peace, friendship and social conversation, to which last Europeans seem also to have applied it in imitation of the savages.

I am informed that they bury their dead like the Creeks with the addition of tying the head down to the knees.

To be particular about the remains of the *Tonicas, Chitimachas,*

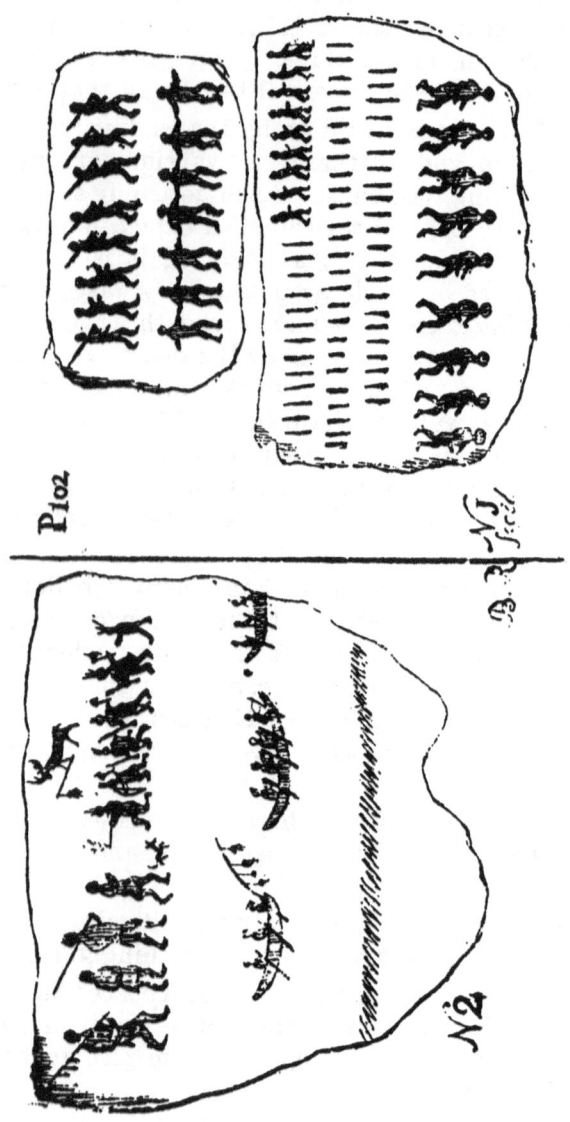

Yasoos, Hoomas, Mobilians, Pasca-Oocoolos, Hattakappas, Taensas, Biloxis, Ybitoopas, Aquelon-Pissas, or *Colla-Pissas, Tiaoux, Oaktashippas,* and many others, who have been destroyed by the French, would be needless, as there are not above from five to eighteen or twenty families of each of them left, who inhabit the banks of the Mississippi, dispersed among the plantations, where they serve as hunters, and for some other laborious uses; something similar to the subdued tribes of New England; and so unnatural is a humanized life to these people, that being as it were out of their element, this remnant melts daily like snow before the Sun.

To give an idea of Indian hieroglyphick painting, I have subjoined the two following cuts: the first Chactaw, and means that an expedition by seventy men, led by seven principal warriors, and eight of inferior rank, had in an action killed nine of their enemies, of which they brought the scalps, and that the place where it was marked was the first publick place in their territories where they arrived with the scalps.

The second is a painting in the Creek taste, it means, that ten of that nation of the Stag family came in three canoes into their enemies country, that six of the party near this place, which was at *Oopah Ullah,* a brook so called on the road to the Chactaws, had met two men, and two women with a dog, that they lay in ambush for them, killed them, and that they all went home with the four scalps; the scalp in the stag's foot implies the honour of the action to the whole family.

Such my reader may be assured were the boasted hieroglyphick paintings of *Muteczuma's* people when *Cortes* landed in *Mexico,* and similar to my description were the people, the palaces and temples of *Mexico* so gloriously painted in Spanish Rhodomontades, for what end I know not, but I can assure every one that these Iberian histories of the conquest of *Mexico* are no other than truth very thickly veiled with absurdities; for except the silver and gold which these savages might then have, believe me they differed nothing from others; and as an evident proof of this we see the free unconquered nations of Darien in possession of gold and silver of their own finding, but very confined in their ideas of working it.

The primum Mobile of the welfare of these countries and of the wealth of their inhabitants are the African slaves; the primary fine spun constitution of Georgia, is a recent and strking proof of this; that country would not this day have been worth the notice of any, had not the importation and use of Negroes been allowed on the arrival of

Governor Reynolds; when this Gentleman came to the Government it had languished under the administration of General Oglethorpe and the other trustees for about twenty years and then the land was worth less than at the first beginning, but no sooner had Mr. Reynolds arrived, and, among other useful regulations, the free use of slaves taken place, the country flourished; it is now about seventeen years ago that this happened: Hutchinson's island before Savannah was then sold for twenty pounds sterling, as yet there is a great deal of it unimproved and it cannot now be bought for ten thousand pound sterling, the improvements likewise are far more perfect; amazing as this increase of value is, yet I believe that there are many instances of a still greater, and I know some lands without any improvements on them of any kind, increase in six years time from five to forty shillings per acre, and many inhabitants of savannah can now testify, that before the surrender of the trustees, almost every house there might have been bought for the consideration of doing the militia duty in the room of the first owner only . Last year the exports of the province amounted to above 121,000 £. sterling, as may be seen by a state of the custom house books then published, an amazing increase in seventeen years from nothing!

The late foolish, not to call them cruel attempts of settling East Florida by whites from Europe (I mean as well from England as from the Levant) are likewise a very absolute conviction of the necessity of having Negroe slaves; but as some people who are able to purchase, slaves, run away with the notion of the unlawfulness of holding a property in Negroes, and who are perhaps not a little misled by the silly pamphlet published in some of the colonies, called "an address to the inhabitants of the British settlements in America upon slave keeping, &c." have attempted to settle without Negroes at all; I cannot in conscience forbear to give my advice to all adventurers in Florida, who desire to improve a plantation for their benefit, not to forget these useful though inferior members of society; not but poor families may live in plenty, and by honest labour acquire a comfortable and easy situation in life as may be wished for, but until their industry helps them to the means of buying one slave and so on till they get more it will be vanity for them to hope for an accumulation of wealth.

Do we not see Solomon's words verified in Negroes? *A servant will not answer though he understand.* The very perverse nature of this black race sems to require the harsh treatment they generally receive, but like all other things, this is carried into the extreme; far be it from me to approve or recommend the vile usage to which this useful part of the creation is subjected by some of our western nabobs, but against the Phyllis of Boston (who is the *Phoenix* of her race) I could bring at least twenty well known instances of the contrary effect of education on this sable generation.

Treachery, theft, stubborness, and idleness, the first in the more northern Negroes, and the three last in the *Ebo, Angola,* and *Benin* slaves are such consequences of their manner of life at home as to put it out of all doubt that these qualities are natural to them and not originated by their state of slavery.

Had *Montesquieu* been well acquainted with the American colonies, he would not have made use of any argument so much below so great a man as the one quoted in the above named enthusiastical production, which seems calculated to procure a greater number of vagabonds than we are already pestered with. I think no man ought to be allowed the manumission of his slave except he be bound for his good be haviour and industry, and idle free blacks ought to be sold for the good of the community.

The anecdotes of the sublimity of Negroes sentiments in their own country are very similar to my anecdote of a savage shooting himself because his mother reprimanded him, or another doing the same because he lost his all at play.

Mr. *Le Poivre* has mentioned the making sugar in *Cochin China,* by free men, but it has not been observed, that those free men make small quantities, and sell it at the low rate they do, only because others who employ slaves in the same or in the adjacent countries, sell them at that price, which may be easily proved, and consequently oblige the others to do the same; nor that these free men live on the same diet as our slaves, and who will say that *Cochin* Chinese sugar after exportation sells at a more reasonable rate than Javan, Martinique, Jamaica, or any eastern or western sugar does.

Have not all the endeavours of the several Legislatures in the West Indies to introduce more white or free people proved abortive, by reason of their general inability for labour in those climates?

Is it not therefore better to employ those, who labour at a similar work in their own sultry country, and in a state of slavery too, than to make victims of men who can by no means be qualified for the fatigue of a southern plantation.

We have known not long ago sugars &c. as cheap in the West

An AGGREGATE and VALUA[TION] from the PROVIN[CE]

With the NUMBER of VESSELS and TO[NNAGE] from the Year 1754, to 1773. Compiled by WIL[LIAM BROWN] Majesty's Customs in the Port of *Savannah*.

	Square rigged Vessels.	Sloops and Schooners.	Vessels.	Tonnage.	Barrels of Rice.	Indigo.	Deer Skins, raw and dressed.	Beaver Skins.	Raw Silk.	Tanned Leather.	Tobacco.	Timber of all kinds.	Staves of all Kinds.	Shingles.
	No.	No.	No.	Tot.	No.	No.	Lbs.	Lbs.	Lbs.	Lbs.	Lbs.	Feet.	No.	No.
From 5th Jan. 1755, to 5th Jan. 1756,	9	43	52	1899	2299	4508	49995	120	438	3250		387849	203225	240690
Jan. 1756, to Jan. 1757,	7	35	42	1799	2997	9335	39220	380	268	6035		289843	196259	263000
Jan. 1757, to Jan. 1758,	11	33	44	1559	2998	18150	26357		358	9837		270396	182268	1784000
Jan. 1758, to Jan. 1759,	4	17	21	665	2371	9600	5791		358	10350		50215	63330	68985
Jan. 1759, to Jan. 1760,	13	35	48	1981	3603	555	7380	622	734	12030		278066	102959	808580
Jan. 1760, to Jan. 1761,	7	30	37	1457	3283	11746	*65765	2298	558	34725		283961	80500	581200
Jan. 1761, to Jan. 1762,	9	36	45	1604	4666	1552	11140	60	332	11775		307690	50969	606650
Jan. 1762, to Jan. 1763,	22	35	57	2784	6509	9133	42855	5260	380	17100		417449	325477	685265
Jan. 1763, to Jan. 1764,	34	58	92	4761	7702	8764	184737	240	953	16605		917384	594356	1470120
Jan. 1764, to Jan. 1765,	36	79	115	5586	9690	14151	172425	540	898	58005		1043535	423251	2061151
Jan. 1765, to Jan. 1766,	54	94	148	7685	12224	16019	200695	1800	711	34575		1879454	661416	3722050
Jan. 1766, to Jan. 1767,	68	86	154	8947	14257	14366	273460	1121	1084	33120		2101466	737898	2036947
Jan. 1767, to Jan. 1768,	62	92	154	8465	11281	12918	205340	5412	671	46670		1767199	748166	2570725
Jan. 1768, to Jan. 1769,	77	109	186	10406	17773	20041	306510	207	541	39234		1782558	806609	3669477
Jan. 1769, to Jan. 1770,	87	94	181	9276	16740	13908	288870	2405	332	33963	3030	1634133	747903	3474588
Jan. 1770, to Jan. 1771,	73	113	186	10514	21129	22336	284840	1469	290	44539	13447	1805992	466276	2896991
Jan. 1771, to Jan. 1772,	64	151	185	9553	25232	19900	270860	528	438	48209	34875	2159075	402253	2247598
Jan. 1772, to Jan. 1773,	84	133	217	11246	23540	11882	213475	632	485	52126	176732	2163582	988791	3525930

* N. B. *The sudden Increase and Decrease of several Articles in certain Years are owing to Difference of Prices, accidental Sh[ortness of] Produce from such seemingly so. The* RICE *from 1772 to 1773 amounted to 96,266 l. 9 s.* DEER SKINS, *raw an[d] of every Year 30 l. upon an Average for a Gain in the Islands of at least 200 per Cent. on Scantling, and 75 per Cent. on P[ine]*

BEFORE me ANTHONY STOKES, Barrister at Law, Chief-Justice of the said Province of Georgia, personally appeared WILLIAM BROWN, Esquire, Comptroller and Searcher of the Port of Savannah, in the said Province, and made Oath, That the above Aggregate is true, according to the best of his Knowledge and Belief; which Oath was taken before me, the first Day of March, 1773.

ANTHONY STOKES.

It is to be observed that the first ninth Year of the above Aggregate amounts to 47,551 l. and the second being 121,677 l. is [in] portion, the Exports must amount to 300,000 l. though much more may be expected if we reckon geometrically; especially if w[e consider the] Acres of Indian Hunting-Grounds, ceded anno 1773, as rich upland Soil as any in America, and exactly adapted for raising G[rain.] America. In the Year 1772 a Gentleman upon the Island of Skid-away, ten Miles below Savannah, cleared 50 l. sterling to so strongly impregnated with Turpentine, resists the Worm and Weather better than the Pine of the more northern Colonies; in C[rops of] forty to fifty Bushels per Acre; and up the Country, from sixty to seventy Bushels; but though there is also great Benefit to be reap[ed from] great Apparatus necessary for that Business. The Floridas may be regarded in the same Light with Georgia, allowing for the

ION of EXPORTS of PRODUCE
E of GEORGIA,
NAGE employed therein, annually diftinguifhed,
AM BROWN, Comptroller and Searcher of his

Turpentine.	Pitch.	Tar.	Pork.	Beef.	Hogs and Shoats.	Corn.	Flour.	Rough Rice.	Peafe.	Sago Powder.	Orange Juice.	Tallow.	Bees and Myrtle Wax.	Hides.	Mules.	Steers and Cows.	Value of fmall Articles not enumerated in the Table.	Total Value in Sterling at current Prices.
Bar.	Bar.	Bar.	Bar.	Bar.	No.	Bufh.	Lbs.	Bufh	Bufh	Lbs.	Gal.	Lbs.	Lbs.	No.	No.	No.	£.	£.
		45	20	40	76	600		237	400				960	48		16	24		15744
			300	126	610	200		40	200			500	150	23		46	32		16776
		129						712					793	1			6		15649
			35	22				100									1		8613
	83	35			700			290	95				100						12694
		425	8	14				208					3910						20852
160		235	274	37		32			271				4584	1050	24	2		5	15870
		246	392	38	70	1280		776	430		384	5120	1800	46			65		27025
8	23	175	161	11	310	405		480	632				1780	65	15		51		47552
19		359	154	78	740	6756		2533	460		316	512	840	179		32	149		55025
		486	394	141	1360	7805		3113	300				100	2170	109		69	318	73426
82	506	723	754	76	965	4527		2892	143	608	536	1280	2051	139		152	184		81122
88	627	387	948	139	777	403		6486	40	100	142	512	3300	132	20	77	66		67092
202	491	167	512	394	1228	10897		1543	440	2912	660	512	3117	270		11	37		92284
68	492	138	673	190	608	21896		1122	273	12289	434	4985	4808	266	20	24	49		86480
103	80	105	521	639	605	13598		7064	601	18405	605	1079	4058	345	30	25	126		99383
45	193	102	409	418	216	11952	7200	1364	96	12930	1770	498	2666	65	21	30	125		106387
40	364	298	628	555	574	11444	1000	2627	140	14435	284		1954	257	10	136	300		121677

Of the above Aggregate the Port of SUNBURY *exported laft Year £. 20,876.*
before or after the New-Year, good and bad Seafons, or Indian Hunts; and Care has been taken to diftinguifh the real
fed, to 18,740 l. 8 s. 11 d. and LUMBER *of all Kinds to 12,634 l. 7 s. 2 d. There is added to the Value*

equal to two and a half Times the firft; fo that in the third ninth Year (anno 1782) by an Increafe in the fame Pre-
der the great Bodies of uncultivated Swamp Lands ftill remaining in the Province, and near four Millions of ungranted
f all Kinds, Tobacco, Hemp and Indigo. In fhort, Georgia may be juftly deemed one of the moft flourifhing Colonies is
Negroe employed in planting Indigo; and there is almoft an equal Profit to be made by cutting Lumber, which, from its being
t of this it is in great Demand in the Weft-Indies.——As to Indian Corn, its common Produce in the Lowlands is from
cultivating Rice, &c. only the wealthier Sort of People can attend to it, on account of the Number of Slaves, and the
e of Infancy.

Indies as in the East; it is the increase of consumption in Europe and North America that renders it dear: I affirm that in America, neither sugar, rice nor indigo can be made by whites at three times the price it is made now by the blacks, and I also affirm that the West Indies lands are distributed among all nations with less reservation, than they are on the continent of British North America.

Can any one say that the favourites of mankind (I mean liberty and property) are any where enjoyed in Africa?

The rhapsodical opinion that the earth produces more when worked by free men than by slaves may do in theory but not in practice; the contrary is easily made to appear; and I am certain from the nature of the climates, that the same colonies when cultivated by free men would not produce one tenth part of what they do now; as for an equal distribution of property is is like Harrington's *Oceana* or Sir Thomas More's *Utopia*.

An European will outlive a Creole by means of his more regular life, not otherwise, and who knows not that Negroes attain with all their labour a much higher age than the generality of idle Whites.

The foolish argument of the shortness of the Jewish history, as well as lugging in the practice of Polygamy by the head and shoulders (although the text quoted is not to the purpose) is too mean to be refuted.

Let not therefore the narrow system of morality adopted by some of our contemporary enthusiastical Philosophers restrain us from properly using this naturally subjected species of mankind; the impossibility of an European's bearing the requisite labour in those climes is now so well ascertained as not to require any elucidation, nor can any one pretend to say, that the posterity of Europeans born in the torrid Zone ought to bear its inclemency; what labour can we expect from men brought up in ease and affluence? It is pretended that our employing slaves is contrary to the precepts of the founder of our most holy religion, when he says: *"Thou shalt love thy neighbour as thyself;"* I need not make use of the confined ideas of the antient Jews, who thought that the title of neighbour did not extend to any thing beyond their own nation.

I will then affirm, that there is a fivefold state of slavery, not only known or permitted; but commanded in God's holy word.

1st. Those who are condemned to slavery for their crimes, which we but too often experience to be the case with the slaves imported to us from Africa.

2ly. Those that are taken in war, which is the most general way among the Negroes to furnish us with slaves, and who would be murdered; did we not induce their conquerors by our manufacturers and money to shew them mercy.

3ly. Those who are sold by their Parents, which custom obtains among many people even the refined and civilized Chinese, not to mention some christians, but most among the Negroes.

4ly. Those who sell themselves or are sold for debts, or other wants, which not only the Negroes, but our own laws justify.

5ly. Those who are born in slavery.

I need not seek assistance from the laws actually in force among us to prove the first; *Noah* in *Genesis*, chap. 9, v. 25, 27, condemns *Ham* to slavery for the crime he had committed, and in *Exodus*, chap. 22, v. 3, God commands us that, "If the Sun be risen upon a thief, it shall unto him be a crime of blood, for he should make full restitution, if he have nothing, let him be sold for his theft." Here let me remind my reader, how much easier the slavery of our Negroes is, than the cruel captivity of those, who for their misdemeanors are condemned to the chain, the wheelbarrow or the gallies.

For the second read *Joshua*, chap. 9, v. 23.

Does not *Exodus*, chap. 21, v. 7, shew, that a man might sell his child as a servant, only giving the maid servant a privilege above men servants?

In the fourth case I think that (without leaving the point in view) I may ask, does not the soldier sell his liberty to his sovereign or other Prince for his pay be it for a time or for life?

Proverbs, chap. 11, v. 29, tells us, "That the fool shall be servant to the wife of heart." Chap. 22, v. 7, "The rich ruleth over the poor and the borrower shall be servant to the lender". Does not Jesus himself, Math. chap. 18, v. 25 make use of a parable fully to my purpose? when he says: "But for as much as he had not to pay, his Lord commanded him to be sold, and his wife and children and all that he had, and payment to be made."

Exodus chap. 21, v. 4, tells us that: "if the master have given his servant a wife, and she have born him sons or daughters, the wife and her children shall be her masters."

For the perpetuity of slavery read *Leviticus* chap. 25, v. 39, to 47, how absurdly is *Montesquieu* again quoted, when in his L. 10, C. 3, he speaks of a conquered nation, who through necessity are made slaves in their own country: every one knows that such conquered people, who are suffered to remain at home at first under oppression, ought by degrees to be made free subjects, their good behaviour being their ransom; but who, even among those refined moralists, would emancipate his slave without some reward either private or publick?

How *Tacitus* comes to be so pat to the purpose of this acute Philosopher I have not been able to learn; nor is the quotation from the universal history, or Mr. Robertson of any more force: Since we are able then to get slaves in a lawful manner, why should we now be restrained from buying them to cultivate our grounds, when all nations at all times have enjoyed that privilege?

Had not the well known Christian Doctor of the Colossian Church a slave called Onesimus? did not this slave run away (after having, as usual slaves, robbed his master) come to Rome and go to see St. Paul? Paul treated him kindly, instructed, converted, and baptised him and sent him back to his master with a letter full of godly eloquence to persuade *Philemon* to forgive his slave and re-establish him in his favour, but by no means an exhortation, much less an order to set him free.

It is not religion then nor christian charity that forbids us to have slaves, but it commands us the duties we are to fulfil towards them, instructing them to obey us, and to use them as a part of the reasonable creation.

Who knows not, that the Spaniards are so much slaves that they can not well be more so, were they bought by their King: And who, that is acquainted in the Spanish dominions, knows not, that a modern Iberian does himself more honour by saying (*Soi Blanco*) i.e. I am a white man, than if he exclaimed, I am Noble!

Shall after what I have said, this Rhapsodist with his confined ideas, send us to some modern system of religion, or say, that I have offered any thing contrary to the sublime doctrine of the author of Christianity? Had this anonymous writer instead of playing with the word slavery, told us, that the Northern colonies had no occasion for Negroes, he would have said more, than all he has advanced in his futile publication.

A Negroe at the Mississippi is reckoned to bring in his master an hundred dollars per annum, besides his share towards all the provision consumed in the family; Negroes in general are used with more lenity there than in Carolina.

I have in my dissertation on the origin of the savages, which has swelled beyond my intention, made mention of a peculiar characteristic of the Negroe species, here let me be allowed to mention that they, like all others of the different species, and varieties of the human genus are born white, which colour soon changes, but on the moment of birth in both sexes the exterior parts of generation will shew, whether the person will be black, yellow, brown, red or any other colour known among mankind.

There is among the Negroes a kind of anomalous beings with white skins, feeble eyes, &c. which have been so often described by other authors that I shall not trouble my reader with a particular account of them; yet one observation I cannot forbear to make, which I believe will be found no less curious than new:

Through all the Northern America the red men are universally so, but as soon as we come on the *Isthmus* we find some of those unhappy anomalous individuals among the savages; I have myself seen several of either sex of this forlorn kind among the *Sant Blass* nation, who call themselves *Ayomalas;* they inhabit the South Eastern part of the *Isthmus,* and I understand from others conversant farther South, that in South America it is common; these people are of such an exceeding tender texture of skin, that it is disagreeably fair and very liable to become scurfy and freckled; their eyes are nearly as red as those of a white rabbit and very weak, seeing objects plainly only after the Sun set, before Sun rise, or in a cloudy day: Their hair is straight, long, lank, and red, not participating of the harsh strong texture of the hair of common savages; their inability for labour, for want of sight, causes the others of the nation to support them; the offspring of these poor creatures however, is again red and in the general course of nature.

The manners and way of life of the white people in Florida, differ very greatly from those in other provinces of America, particularly in respect of cloathing; they are very plain, their dress consists of a slight waistcoast of striped cotton, and a pair of trousers of the same, and often no coat; if any, it is a short one of some light stuff; in winter a kind of surtout, made of a blanket, and a pair of Indian boots is all the addition; the women also dress light and are not very expensive; happy frugality! May the inhabitants of this blessed climate long continue to cherish thee as their greatest temporal blessing! Manufactures of cotton for their own use also prevail greatly among the industrious Acadians; may the new comer under the free government of the English not be ashamed to follow this excellent example! sorry I am though, that I find myself obliged to say that when in anno 1772 I was in Orleans I could not forbear to feel the weight and justice of an exclamation made by a French gentleman to me upon seeing some English gentlemen walk upon the Levis "*Dites moy Monsieur (says he) lesquels sont les petits maitres de la Louisiane Messrs. les Anglois ou nous autres, with grief I was obliged tacitly to own the English were.

The amazing plenty of the country in its western regions makes them keep princely tables at a small expence, thus I can not call this luxury a fault; to form an idea at what a good table may be furnished, I beg my reader to peruse the following account, as this really stood in the year 1772 in the months of November and December, and in January and February 1773: Beef at 13s 6. per hundred; fresh pork of the best corn fed 18s per hundred; country flour from 12s to 13s 6. the North American flour 15s 6. to 16s 6. per hundred; a bushel of good clean full grained Indian corn about 13d. ½; a barrel of the best clean merchantable rice being about two gross hundred weight of neat grain, from 9s to 11s 3; sweet potatoes about 8 d. per bushel; a sheep of about 12 pound per quarter, if purchased above New Orleans, will cost 6s 9, if below, 9s; a tame goose of the best kind from 10d to 1s, a turkey about 20d. to 2s; if a very fine wild one, some times 2s 6; a common dunghill fowl about 6d; a tame duck from 6d to 8d; venison variable according to the caprice or nation of the hunter, but always very cheap, this has a noble peculiar fine flavour here; butter is little used, yet the best the Acadians make which is very good sells at about 6d. ¾; bears oil carefully clarified supplies its place; this is bought at about 3s 6 a gallon; and in my opinion, as well as in the opinion of all who once try it, richly deserves the preeminence of butter; besides these the lakes Pont-Chartrain, Maurepas, Borgne, and Ouachas supply the country with a vast variety of excellent fish at an exceeding low price, and in winter large quantities of fine oysters and a great choice of water game in high perfection, so low as seldom to be asked more than 6d. for a goose, about 4d. for a duck, mallard, widgeon, or other of that size, and about 2d. for a teal;† all these have here a

---

*Tell me Sir, which of us are the Petits Maitres of Louisiana the English or we.

†All these prices are as near as possible reduced out of the country currency into sterling.

peculiar flavor and are much admired; the river abounds likewise in fish, but being very deep and little pains taken to catch them, the different kinds are scarcely known; the planters all keep their hunters for providing their tables, and their yards abound in poultry; small game, such as quails, pheasants, hares, squirrels, snipes, godwits, curlews, meadow larks, fieldfares, rice birds, &c. &c. are very frequently had.

It is true that *Pensacola* and *Mobile* especially the first are not so abundant; but in regard of meat, venison and fish, I pronounce it to be owing to the indolence of the inhabitants, who content themselves to pay any price fixed thereon by one or two butchers, and three or four industrious Spanish hunters and fishermen.

Mobile has plenty of bread kind already, and notwithstanding the barren miserable appearance of the sea coast here and at *Pensacola*, their fine rivers will e're long oblige propitious plenty to rear her at present hidden head.

In respect of vegetables, from *Mobile* westward there is an amazing supply of every kind the year round in their highest perfection and of the finest flavour; but the barren sand of *Pensacola* will not admit of gardening in June, July, and August, the proportion of moisture being then too small for that of heat.

In the gardens we find various kinds of cabbage and sallading, carrots, turnips, radishes, skirrets, leeks, scallions and other roots, asparagus, artichoakes, cucumbers, green pease (in some the year round) vast variety of beans, and choice of best flavoured pot herbs, in short every fine esculent plant Europe can boast of, they have beyond description perfect, neither has the bounteous hand of nature here been sparing in the most delicious fruits for the table, nor forgot to stock the fields and woods with fine mushrooms, truffles and morels for the most exquisite sauces.

Hic ver assiduum atque alienis mensibus aestas bis gravidae pecudes; bis pomis utilis arbos

VIRG.

Happy climate where through all seasons of the year the inestimable gifts of *Flora* and *Pomona* are common, where snow or ice are very seldom seen, and where the cruel necessity of roasting one's self before a fire is utterly unknown!

With respect to drink, among the English inhabitants it is generally water tempered with a moderate quantity of the best West

India rum, and among the better sort the Portuguese and Spanish wines; the French drink their favorite claret of the two kinds called *vin cahor* and *vin de ville;* the Spaniards add the St. Lucar wine to it, New England rum (that bane of health and happiness) has found its way here also, among the lower classes; a low kind of rum from the French Isles is likewise used by some, and its cheapness makes it preferred for the trade with the savages.

A transition to the agriculture of these colonies comes most naturally in course; this science (undoubtedly the most noble and most useful of all employments) has not only had the protection and countenance of Monarchs and other great men, but they have even practised it, our present gracious Sovereign has shewn instances of his care for the art, and many individuals of rank in England, Holland and France now practise it and study its improvement, and the annals of all nations tell us, that the flourishing state of that art was the happy period of their grandeur; and it undoubtedly is and always will be an honourable, innocent, pleasant, and usefull pursuit.

The amazing multiplication of mankind in America, which since eighteen years that I have been acquainted with the continent has proceeded nearly, if not wholly, in a triplicate proportion, is entirely owing to this great art, and to the room this wide waste of new world affords for its improvement; this art is the source of our grandeur, and the cause of our happy strides to greatness; the welfare of the people depends thereon, all the consequence of a nation is owing to it, and the national independency of the greatest communities is in a great measure inseparable from it.

In this noble country then (which will afford not only all the necessaries but even the superfluities of life) it ought therefore to be the principal object of our most arduous pursuits; all the products of the torrid Zone as well as of the temperate are capable of being produced in this perfect climate, let us see which of them most deserves our attention.

Before I descend to particulars a thought claims my notice; West Florida is beyond measure happy in having its fertile soil more equally divided than the barren sands which were chosen by the* monopolizers of East Florida; these even overlooked the most useful places there, and planted their baronies in the pine barrens. There

---

*No doubt this was a scheme of the enemies of American population who see themselves every way baffled.

let the lords be lumber cutters! The fertile part of West Florida is more equally divided, and there population through agriculture does and must continue to flourish, happy circumstance! for which Dr. Stork and the self taught Pennsylvanian Philosopher deserve our perpetual thanks.

1. Wheat will always be produced in West Florida especially in the higher latitudes on the river, so that all attempts to starve that country by blockade would be vain and frustrated; barley, and oats grow well in all the lighter soils, and rye may be produced in the most barren sands of the province.

The cultivation of these are so universally known, that to describe them would be spending paper in vain, but a remark in regard to wheat (which has fallen under my observation) must not be omitted; in the lower latitudes wheat will grow exceeding well till it comes almost to maturity; but then of a sudden a disease called the blast will often take it, and this in one night reduces the hopes of a fine crop into a certainty that the best thing to be done next morning is to burn the straw to clear the ground for some thing else; to prevent this an observant farmer of my acquaintance in Georgia* having found that rye was never subject to this disease, mixed some rye among his wheat, and the field escaped the blast; he repeated the trial till he was convinced of its efficacy and then sowed a field of about five acres with wheat, surrounding it with a list of about twenty feet of rye; he succeeded in this experiment also, and his repeated trials have now shewn that it may be depended upon as efficacious, and thus wheat may be truly made an universal grain.

II. The *Zea*, Maize or Indian corn is a grain by no means to be neglected; the rich grounds in the N.W. parts of East Florida about *Alachua* &c. and the lands in the N.E. of West Florida are peculiarly adapted thereto and will yield fifty or sixty bushels per acre, the intervale on the *Tombechbè* and the land on the Mississippi is capable of producing eighty bushels per acre; even the sandy land, or that which is little better than sand, will yield according to seasons from fifteen to twenty five bushels an acre.

The vast consumption of this grain in food for mankind is very well known, but its superior goodness for the feeding of swine, poultry, and indeed all animals, joined to its easy culture and prodigious increase, makes it deserve a rank before all other grain after

---
*Mr. Isaac Young.

wheat: Its culture requires the ground to be thoroughly hoed* or plowed, so as to make it entirely mellow, and break every clod, then holes are made at equal distances in regular rows, and four or five grains dropped in each and covered; in this part of the husbandry the greatest mistake is often made through over greediness of land, the holes ought to be at the regular distance of five feet, which will leave room for the horse or hand hoeing it much better, than if it was placed at four feet; some people are so near minded as to plant them only three feet a part, but this prevents the free circulation of air through the rows, of which this grain stands very much in need; this is so evident, that on planting in the same field some corn at three, at four, or at five feet distance, (suppose the soil equally good) there will be a difference of six feet in the height of the plants, for the best land can not raise the crowded corn to above ten or twelve feet, whereas that through which the air has free access, is frequently raised to sixteen, eighteen, and even twenty feet;† nor is this the only difference, the crowded corn will not produce grain in any thing like the proportion of that which stands at proper distances; a stem of crowded corn will hardly ever furnish above three ears, and these not above eight inches long, nor have they their extremities ever filled, neither are the rows so compact nor close together as that of corn that has enjoyed more freedom; a stem of which will produce generally from three to five well grown ears from nine to twelve inches long, compact, and to the end full grown and of a wholesome look; when the young plants are grown about five or six inches out of the ground, a man ought to go through the field, and pull up those plants that look least promising leaving only three plants in each hill and the ground must be kept weeded; when at the height of a foot or thereabouts, it must have the first hoeing, the earth is then to be drawn round upon the root in form of a cucumber hill; and now it may stand till it attains the height of three feet, when it must have a second hoeing, and be cleared of suckers, which will appear in numbers; when the grain is got to the growth of about six or seven feet, or when the tassel or female flower begins to appear, it must again be relieved from suckers, and have a third hoeing, throwing the ground well up round its foot; in a large field these hoeings are most commodiously performed by the hoeplow drawn

---
*The savages anciently performed this with the shoulder blade of animals fixed on staves.
†I write for Florida.

by one horse, but in new fields the hand hoe must assist, if not do all the work; all the trouble with which this culture seems burthened is amply repaid by this grateful grain, the time of ripening depends on the climate; the northern climes bring it tardily but the southern ones hastily.

After this third hoeing the spike or male flower will appear in perfection; you may now strip some superfluous leaves from the lower part of the plant which will prove excellent hay for a horse and some cows that are necessary to keep near the house; the corn having attained nearly its growth and the spikes not shewing any more *pollen*, which is the impregnating yellow dust, that is seen on it during its flowering state, we find the female flower or silken tassel change colour; the ears fill and some lower leaves begin to fade, it will not be bad husbandry to cut off all the tops above the uppermost ear; this is left two or three days to dry on the summit of the plant, then tied in sheaves, it is as fine a hay as any known for feeding such animals as are intended to be kept about the farm yard, or in the stable, and every prudent planter will find it his interest to keep some milch cows (at least) at home; this lopping of the tops is also said to contribute to the filling of the ears and the sooner ripening of them: the corn ripens in four or five months from the planting, in Florida; the gathering is done by pulling the ear from the stem, and carrying it in baskets to a place where it is stript of all its husk except the inner coat, and then it is kept in some airy loft or granary: and when wanted for use, it is laid in heaps and threshed with flails, and afterwards winnowed like other grain.

A report has been spread that on the Mississippi it was only necessary to burn the canes, make a hole in the ground and put the grain in, and it will grow without culture; it is true, but less wise men than Solomon will know such a field at sight to be that of the sluggard, and know the owner by his shabby appearance.

To enumerate the vast variety of ways in employing this noble grain for food such as *hommany*, mush, groats, parched flour, cold flour, puddings, *Anagreeta** and a multitude of other dishes, that vary in the difference of ripeness, grinding or other ways of dressing it, would be too tedious and unfit for this work; however the mixing

---

*Anagreeta is the corn gathered before maturity, and dried in an oven or the hot sun by which means it retains its sweetness and is easily dressed, making a fine mixture in puddings especially with pease, but this is only practised in the provinces of New York and New Jersey.

its flour in equal parts with wheat flour after macerating the first in water till the liquor becomes slightly acidulated makes so fine a bread that I could by no means think it just to forget the mention of this process, the meal coarsly ground and cleared of bran, then boiled with a little salt in water, and thus mixed with wheat flour, makes also a palatable bread of great use to the poor; I hope my reader will excuse this prolixity, so great a staff of life deserves some notice.

III. Pease, as they are here called but improperly, because species of the *Phaseolus* and *Dolichos* are meant, follow the maize in utility: It is well known that most people use them like European pease either green or dry, and some kinds, such as the small white sort, the bonavist, cuckolds increase, the white black eyed pea, the white crowder, and many others, are undoubtedly at least as good; add to this that while young, hull and all, they make a fine esculent dish for the table, and when hulled they are as good as green pease, and as much admired; the hulls after threshing are eagerly sought after by cattle, and increase milk, the hogs fattened with this pulse are the next best pork to those fed with maize; thus they infinitely increase the quantity of food; their culture is easy, they are generally now planted between the corn at the second time of hoing; they want little or no attendance in that case, as the corn serves them for support to climb up by, and the further attendance on the corn also serves the crop of pease; I can not but think this husbandry a very good one as by the time that the *Cirrhi* take hold of the corn it is sufficiently tilled to be out of all danger of hurt from this parasitical nature of the pease. I also believe that the haulm left behind supplies the land with sufficient manure to re-establish its vegetative vigour, which maize is but too apt to exhaust.

The proper pea is not so fit for the field in this part of America, therefore only cultivated in gardens for the purpose of eating them green.

IV. The esculent *convolvulus, vulgo,* sweet potatoe, claims the next place in this list; this root is by agriculture meliorated so much that some of the varieties are by no means inferior to any food we use on the southern tables, although some palates, for instance my own, can hardly accommodate themselves to them; the following list will point out the varieties, in an ascending scale for goodness.

1st. *Spanish,* or the original root.

2d. *Carolina,* little superior to the first.

3d. *Brimstone,* from its internal colour, with a red skin.

4th. *Purple potatoe,* having that colour throughout except a very little of the heart.

5th. *Bermudas,* or round white potatoe.

The 1st. 4th. and a few of the 5th. are cultivated in the Floridas.

The 1st. is scarce fit for the table being very fibrous, therefore most proper to feed cattle; however pork from hogs fed with them is indifferent, and requires to be hardened a considerable time with corn; it is remarkable that in pork fed with them the fat always separates wholly from the lean, which is likewise the case with that fed on the common peruvian potatoe, vulgarly called the Irish.

The 4th and 5th are excellent food and deserve a place on every table; the 4th cut into longitudinal slices and fryed is a very good dish; plainly boiled they are a good *succedaneum* for bread.

The 5th being less sweet and more dry than the others are best for stewing with meat, such as fat pork or beef, or a fat goose or duck, to make what is called an *Haricot*; their very mealy texture renders them the most proper in room of bread, or to mix with flour and make bread of.

They are a profitable crop, and require a light sandy soil which must be made very clean and mellow; they are planted in beds or hills, being propagated from pieces that have what is called an *eye* in them; they require two or three hoeings, and with this management will produce from three hundred to five hundred bushels per acre; even the last, if we reckon ten hills necessary to make a bushel.

About July, in rainy weather, slips are taken from them, and planted in beds to procure a crop of small ones for next years feed.

The very same treatment is here necessary for the peruvian potatoe, but it wants oftener covering because the heat of the summer sun would strengthen the poisonous juices (with which this genus of night shade abounds) in those that might be exposed to the air; therefore they are unfit for the field in this climate, nor will they bear to be kept any time but in the garden; they will yield six or eight crops yearly, of a very good kind for the table.

V. Buckwheat justly deserves the next rank, it being the most fattening grain to all animals, but especially hogs and poultry; which last are always surprisingly multiplied where this grain is raised;

to man it is also an excellent food; it is well known that in *Philadelphia* buckwheat cakes are one of the articles of that city at their breakfasts; it is also a noble crop near an apiary, and will multiply honey greatly; it requires a light loamy soil well broke, and to be sowed very thin, it improves land where ever it is planted.

VI. The *panicum* or Guinea corn is next in this scale of the produce of the earth; it differs from maize in being more difficult to be reduced into food, and being of too hot a nature for brutes, especially poultry, who will become blind by eating it often; it impoverishes land, but when sown at broad cast will yield a fine and profitable crop of hay for such as are inclined to keep horses or milch cows near home, nor has it in this case so bad an effect on the soil.

VII. Rice though so lucrative a commodity did not deserve an earlier place in this enumeration, because it is not properly a necessary article in the human oeconomy in this country but a staple branch of trade, not likely to increase the number of profitable members of society; however notwithstanding it is (for our use) only fit for puddings, and to put in soops, or to make the wafer-like bread called journey cakes in Carolina, yet it must be mentioned here on account of its usefulness in feeding Negroes, cattle and poultry.

Its culture is in wet grounds, where in the dry season it is sowed in drills eighteen inches apart and kept very clean, first by hoeing, and afterwards by letting water on just sufficient to destroy the weeds, and not to drown the rice by covering its top; for rice will grow in any soil though it loves watery ones best; but the reason of letting water on is more to suppress other plants than to forward the growth of rice; for the crop grass and other weeds are kept under by being covered; cover the rice and you will also destroy it; and I have seen good crops in high lands, when the season was but moderately wet; it is therefore to save labour that the planter chuses wet ground, but should he not have the water at his command, the noxious growth in dry weather, or a profusion of water in wet weather, would equally destroy the crop; it is therefore necessary that great banks and deep ditches should be made to secure the crop, and that is no work for new planters to begin.

The grain in the rough is very noble food for all quadrupeds and poultry; to manufacture it there is a great deal of labour necessary; when manufactured a quart per *diem* will very plentifully feed a working Negroe, and the dust, beaten hulk &c. that comes off will cause all animals on the plantation to be very fat, especially turkeys, which no place on earth furnishes equal to those on a rice plantation; it will require eighteen or twenty bushels of rough grain to make a manufactured barrel of 500 pounds weight: The time of planting is from the end of the frost to the 10th of June, and the season for reaping is generally in October; a second crop will rise fit for fodder. In the twelve mile swamp, in *East Florida* I have seen a second crop come to perfection, and no doubt in the grand marsh in latitude 25, 30, in that province it would always yield twice.

An acre will produce from sixteen to eighteen hundred pound weight, manufactured grain, and one Negroe will attend three acres in a very compleat manner.

This grain causes a great consumption of cypress and pine wood to be made into casks; consequently a great deal of coopers labour is necessary, in which the planter ought to be obliged to employ white men, it not being harder labour than is consistent with our temperament and constitution.

The process of manufacturing rice in a great measure resembles that of peeling barley; the planter who has water at command manufactures by a water mill, consequently cheaper than he who is obliged to do it by horses or by the manual labour of his slaves.

This grain has one great article in its favour, that is the number of hands it employs in transporting and exporting it, who may be truly said to live by it:

The machines for manufacturing it have been so often described and figured, that they hardly ought to be introduced into a work that bears the name of concise.

VIII. There is no doubt but *beans* would be found of use, were they cultivated on the strong clay grounds which are at some distance from the Mississippi, and such as the lands on the ridges of the *Yasoo*, and in the Chicasaw nation; but as long as the canes, reeds, and the numberless natural grasses are in such abundance, this article is not yet worth attending to, any more than crops of turnips, carrots, or cabbage, which however would undoubtedly answer here for culture in the field as well as in any other part of the earth.

IX. The artificial *grasses* found here are 1st: That kind of grass known in the islands by the name of dog grass, and in Carolina and Georgia by that of crop grass, and by the French of Florida *chiendent*; this so greatly resembles the *annual meadow grass*, that I would rank it as one of that class; its leaves are broader; this rises naturally after the ground is broke up and will continue in the ground for two or three years, when the ground being again plowed up, a fresh crop ensues, and this is all the culture it wants; in Carolina and Georgia, this is apt to plague the planter much by reason of its continual renovation, and the strong matting of its roots whereby it renders the surface of the earth in a great measure impenetrable to the hoe, wherefore they wish to destroy it; but in places where pasture is scarce and eagerly desired, they would do well to consider its excellent quality in fattening cattle, and the fondness all cattle shew for it; and if instead of trying clover and other foreign grasses, which have so often baffled their utmost efforts, they would encourage this easily cultivated grass by plowing their poorest grounds once in two years, they would soon find themselves amply repaid by their long wished for *desideratum*, pasture; this grass will grow kindly in the poorest soils of Florida.

2dly, A second species of the same kind with narrow leaves called by the French *Herbe au cheval*, from the fondness of horses for it; in Carolina it is called *nutt grass* from a nutt found at its root; this, when once it takes in the ground, is as easily entertained as the first and makes a very good pasture on poor grounds, but it must be well fenced against hogs, which being very fond of the nutts would root all up in a short time.

It may be said that I describe these grasses as natural yet call them artificial, but when we reflect, that they require some kind of culture, the name of artificial can not be thought improper.

3dly, A grass not ill resembling *silver hair grass*, which is called from the colour of its seed, black seed grass, makes an excellent pasture on the meanest sand we find in the country; the seed is easily procured in Carolina, Georgia and East Florida.

4thly, *Scud grass* vulgarly called Scots grass is a noble grass on poor land, it grows to the height of thirty inches and upwards;* an experiment made by Mr. *Wegg* at Pensacola convinces me that this grass will grow in the sandy land of the provinces; the seed or plant may be procured at Jamaica where I knew a gentleman of the name

---

*In wet lands in Jamaica it will grow four or five feet high though the soil be sandy. It is propagated by cutting it in pieces, leaving a joint to each piece, this stuck into wet swampy ground soon grows and propagates others, so that it becomes a very close matted sod; I have seen it called *gramen panicum*.

of *Jones* make 1500 £. per annum of a penn by means of this grass alone: Cattle prefer it to every other.

5thly, The *Panicum* and maize are of use as grasses I have already shewn how.

6thly, A species of *Dolichos* lately introduced into Georgia from *China* although not properly a grass, yet as it thrives to admiration there and yields four or five crops per annum, I think it not improper to recommend, as deserving cultivation for feeding cattle, the more so as all kinds are fond of it.

7thly, Experience has taught us that *sain—foin, lucerne, clover* and *timothy grass* thrive not in the eastern part of these provinces, the sun being too hot for them in summer, at which season clover in particular can not resist it out of the shade; on the western part of West Florida I believe they might answer, but we need not yet look out for pasture there, while that noble grass *Indian reed (canna)* is so abundant.

X. *Sago* might be here produced as well as in Georgia, for the tree from which the basis of this drug is taken abounds particularly in East Florida; every body knows of what a vast use it is.

XI. *Sesamen* or oily grain, This was introduced by some of the Negroes from the coast of Africa, into Carolina, and is the best thing yet known for extracting a fine esculent oil; it will grow in any sandy ground, even luxuriantly, and yields more oil than any thing we have as yet any knowledge of: Capt. P. M'Kay of Sunbury in Georgia, told me that a quantity of this seed sent to Philadelphia, yielded him twelve quarts per bushel; incredible as this may appear, I have the greatest reason to believe him; the first run of this oil is always transparent, the second *expression,* which is procured by the addition of hot water, is muddy, but on standing it will deposit a white sediment, and become as limpid as the first; this oil is at first of a slightly pungent taste, but soon loses that and will never grow rancid even if left exposed to the air; the Negroes use it as food either raw, toasted, or boiled in their soups and are very fond of it, they call it *Benni.*

All the culture it requires is to be sown in drills about eighteen inches apart and by frequent hoeings to be kept clean.

XII. The *ground nut* also introduced by the Blacks from *Guinea,* is next after this for its easy cultivation, a good kind of oil that does not soon grow rancid, and the great quantity it yields; but the earth does not produce the seed in such plenty as he last, and it takes up more room.

XIII. The *pumpkins* as cultivated in the fields here, being of an easy culture in the poorest soils, and yielding a great and beneficial increase of food, should not be forgot, although on account of their being chiefly used as a sauce I have given them this late place; their culture is so easy as to require little or no attendance after the seed is in the ground; they overgrow every kind of grass or weed, and are generally planted by dropping some seeds in the potatoe or corn fields, and their increase is immense; was the shield shaped squash from the north added to this, it would prove a beneficial addition; all these kinds are eagerly eaten by horses and cattle of every sort, and they increase milk.

XIV. Liquor is as necessary as victuals; I would therefore recommend the culture of *vines*, which will here succeed with certainty; ask the French inhabitants and they will tell you, that they once were in a very fair way to make their own wines at least, until an order for the suppression of the vineyards came from France; the remains of these vineyards in many places yet shew the practicability of this scheme; the Spaniards continue the proscription, but why should not we make this profitable use of a country, whose soil and climate are evidently inviting us to this attempt, and which experience has taught us are adapted to it?

XV. *Apples* and *pears* are here of very good quality, but are never likely to become an object of attention by growing in quantities sufficient to make cyder and perry; but peaches grow here of the finest flavour, and in the highest perfection, on standard trees, and therefore are fit to be planted in orchards. It is well known that hogs fattened by them make an excellent pork; the superfluous quantity would no be ill employed in that way, but as in Virginia they have set us the example, why can we not in Florida also distill their juices, and by means of that spirit which becomes excellent by age, at least partially banish the money draining useless article rum? This tree should be grafted not so much on account of the choice of fruit (Florida produces no indifferent ones) as because the tree in this climate, especially in sandy soils, is not so lasting when raised from the nut as when grafted on its own or any other proper stock.

XVI. *Sugar* is a matter at present of mere speculation, yet it is made already near *Orleans*, although acknowledged not to yield the profit of a rice or indigo plantation. At or near *New Smyrna* some is also produced, and rum has been made there. Some of the lands between the latitude 25 and 27, would undoubtedly yield it to advantage particularly at the river *Rattones* and the grand marsh.

We gather from *Horace* and *Virgil* that severe frosts were common in their time at *Rome*; now ice is scarce known naturally there; so while I have been acquainted with New York government, which is about seventeen years, it seems to have altered its climate four or five degrees more southward;* of this we have many striking proofs; this being owing to the opening of the country it is not unlikely but in time, by clearing the woods of Florida, that country may be brought to produce sugar to advantage, till then its culture merits no place here.

XVII. *Oranges* of various kinds are worth notice as they are on many accounts usefull in drink and sauces, and their leaves a good fodder for some esculent animals, such as sheep, rabbits and goats; they thrive extremely well throughout all Florida.

XVIII. *Olives* are as yet a matter of speculation, but as the wild olive is found here, and the cultivated one has already shown its propensity to a naturalization in this country, I make no doubt but they will become a grand article here; their utility is so well known that it requires no comment; their culture can not yet claim room in this work.

XIX. †Hops grow spontaneously through all this country.

XX. Having now gone through most of the different esculent productions of the earth, I will examine the merits of those which are

---

*The later appearance of frost and the earlier arrival of spring, but above all, the visits paid of late years by some species of southern fishes to New York harbour, such as the *mullet, porgy* and some others are evident proofs of this.

†In some of the Swedish provinces, a strong kind of cloth is said to be prepared from hop stalks; and in the transactions of the swedish academy for the year 1750 there is an account of an experiment made in consequence of that report. Of the stalks, gathered in autumn, about as much was taken as equalled in bulk a quantity of flax, that would have produced a pound after preparation. The stalks were put into water and kept covered therewith during the winter: in March they were taken out, dried in a stove, and dressed as flax. The prepared filaments weighed very nearly a pound, and proved fine, soft and white; they were spun and woven into six ells of fine strong cloth. The author Mr. *Schissler* observes, that hop stalks take much longer time to rot than flax; and that if not fully rotted, the woody part will not separate, and the cloth will neither prove white nor fine.

Dr. Lewis's notes on Neuman's chymistry 4to, London 1759, page 429.

serviceable in commerce, be they for manufactures or otherways; and among these *indigo* justly calls for the first rank.

The description of this plant will more properly appear in the botanical part of the work.

For its culture it requires a middling rich loose soil, and the field ought to be as nearly as possible a perfect level; it will grow in any soil from the heaviest to the lightest, but rich hammock, or oak land, of a moist nature, is the best adapted to this purpose; the ground should be thoroughly cleaned, and reduced to a perfect garden mould; this is the most laborious part of the culture, and so absolutely necessary that no crops can be expected without it.

Seed of the best kind abounds on the Mississippi, about four bushels of seed are requisite for an acre, it must be sown in drills about two feet apart; the time of approaching rain is always best; the season for sowing sets in the beginning of March and may be continued down till May; if the season is any thing favourable it will afford five cuttings between March and November; seven weeks being a long allowance between each two cuttings; we must be very cautious about cutting, for if that be done in dry weather it will infallibly destroy the plant; but in rainy weather there is no manner of risque of this; by this treatment and care, the plant is continued for years together in the warmer climates; it ought to be cut as soon as there is any appearance of blossom, ten weeks from planting will generally ripen the seed perfectly; when cut, it is tied in bundles and carried to the vats.

The vats are three in number and ought to be, the first very large, the second one third less, and the third yet less: At the head of the large vat stands a pump to fill it with water; these vats particularly the first or steeping vat, ought to be made of very hard timber; in this steeping vat the weed is thrown together, and pressed down with pieces of live oak or other solid and ponderous timber; it is then covered with water by means of the pump; here it remains to ferment; the *crisis* whereby to know the exact time it is to remain in this vat is when the liquor thickens, begins violently to effervesce and assumes a purplish blue colour; this will be effected in a greater or less space of time from eight to twenty hours according to the temperature of the atmosphere.

The steeping vat projects with one edge about three feet over the second or beating vat; in this edge the bottom of the first has a hole and plug in it, this plug must be drawn as soon as the above signs of the perfection of fermentation appear, to draw off the liquor from the weed, which last is absolutely useless; except perhaps it might be employed to good purpose in a saltpetre manufacture.

In this second or beating vat as soon as the liquor is in, it must be beat or stirred by a process similar to churning; this is a laborious work and used to be performed by Negroes, who draw up and down a lever that has either one or two bottomless square buckets at each end; but of late horses have been employed in large works; this churning is continued till the dying particles are separated from the liquor, or as it were sufficiently congealed to form a body or mass; here lies the secret of the art, for if the beating is ceased too soon a part of the dying matter remains undissolved, and if beat too long some part will again dissolve; only experience can teach this *criterion*, and there is only one method to try it which is by taking up some of the liquor in a phial or cup and observing whether the dying matter is inclined to depose itself or not; all farther theoretical lessons are in vain, the young planter must have recourse to practice.

Lime water is in the English colonies used to hasten the separation, this I am inclined to believe spoils the indigo, neither the French, Dutch, nor Spaniards use any in their plantations.

The indigo being arrived at this crisis the churning ceases, and it is left to subside at the bottom of the vat; when the liquor begins to look of a faint green transparent colour the water must be drawn off, first by a cock fixed at a certain height in the side of this second vat, till you come near to the superficies of the *Residuum* which is the indigo; then another cock corresponding with the third vat must be opened to let the *Residuum* run into this last vat, where it remains to settle a little longer in order that it may totally discharge itself of all the tinging matter,* it is then put into bags of the form of *Hippocrates's* sleave to drain it from all superfluous humidity; these bags must hang in the shade.

When all the water is drained from it, the remainder, which has all the appearance of mud, is put into very shallow boxes, where it is left to dry; when it begins to have the consistence of clay fit to

---

*This will be compleatly deposited in about 8 or 10 hours time, the Residuum must be strained through a horse hair sieve, previous to its being put into bags in order to have it entirely pure and free from extraneous matter.

make brick with, it must be cut with a very thin bladed knife into square pieces, and then further left to become quite dry, which is the state in which indigo comes to us.

This last process must be all done under a shed where the air has free access but the sun none; should the sun touch indigo in this state, it would exhale all the tinging matter and leave the mass in a colourless state, similar to slate in appearance; beware also of moisture for that will keep it dissolved and incline it to putrefacton.

Some planters press their bags in a box of about six feet long three feet wide and two deep, having holes in the bottom to let the water off, and a strong thick board fitting exactly in it; in this box the indigo bags are laid and the board with a number of weights on it, but whether this method is better than hanging them in a shed to dry I know not.

I should have observed, that in the drying shed the pieces must be carefully turned three or four times a day, and that two young Negroes with a bush, wing, or bunch of feathers, ought to be employed in fanning the flies out of the drying shed, as they are hurtful to indigo; be cautious also in packing it in barrels not to put it in till it is thoroughly dryed.

The dimensions of a set of vats in Carolina is about sixteen feet square, and three feet deep, in the clear, for the steeper; and the battery twelve feet square and four and a half feet deep for every seven acres of indigo; they make them of two and a half inch plank of cypress, and the joints or studs of live oak; to these the planks are well secured by seven inch spikes; such a set will last seven or eight years.

The best indigo is called *flotant* or *flora*; this is light, pure and approaching to hard; it floats on water, is easily inflammable, and is almost totally consumed by fire; the colour is a fine dark blue inclining to violet, and by rubbing it with the nail it assumes the colour of old copper.

The next best is more ponderous, and is called violet or *gorge de pigeon*, its colour being alluded to; these two are best for dying or staining linen and cotton.

The third kind is of a copper colour deriving its name from the coppery appearance it exhibits on being broke; this is the weightiest of all the merchantable indigo, and therefore the *desideratum* of the planters; and is most used for the woollen manufacture.

The inferior sorts are not worth describing as they are unsaleable and unfit for use; they discover themselves by flintiness, or a muddy soft crumbling appearance accompanied by a dull blue colour, often appearing even like slate.

An indigo work should always be remote from the dwelling house on account of the disagreeable *effluvia* of the rotten weed and the quantity of flies it draws; by which means it is also scarce possible to keep any animal on an indigo plantation in any tolerable case, the fly being so troublesome, that even poultry thrive but little where indigo is made; nor is there scarce a possibility to live in a house nearer than a quarter of a mile to the vats; the stench at the work is likewise horrid: This is certainly a great inconvenience, but it is the only one this profitable business is subject to.

XXI. *Cotton* being so very useful a commodity that scarce any other exceeds it, and an article of which we can never raise too much (for like all other things, the more it is multiplied the more its consumption increases) it therefore behooves me to mention it as second in rank: We, by following the example of the industrious Acadians, will do well to manufacture all our necessary clothing in Florida of this staple, and although it has not yet been raised in a sufficient extent to export a considerable quantity thereof, yet when we consider the number of manufactures in Lancashire, Derbyshire, and Cheshire that consume this benificial commodity either alone or in mixture with silk, wool, flax, &c. and that England imports all the rough materials from abroad (chiefly from the Levant) to so great an amount as near 400,000 £. sterling value, we may perhaps find it worthy of a more universal propagation; I shall, in hopes that this may take place, give the method of culture and of cleaning this produce of our earth.

The natural history of the plant and its characters, species and varieties will be found in the botanical account.

Cotton will grow in any soil, even the most meagre and barren sand we can find.

The sort we must cultivate here is the *Gossypium Anniversarium* or *Xylon Herbaceum*; also known by the name of green seeded cotton, which grows about four or five feet in height. Give this plant a dry soil and further it will cost you little trouble or attention; it must be planted in rows at regular distances about six feet apart; plant the seed in rainy weather and in about five months time the fibres will be compleatly formed and the pods fit to gather, which will be known by their being compleatly expanded; it must now be carried to the mill of which take the following description.

It is a strong frame of four studs, each about four feet high and joined above and below by strong transverse pieces; across this are placed two round well polished iron spindles, having a small groove through their whole length, and by means of treddles are by the workman's foot put in directly opposite motions to each other; the workman sits before the frame having a thin board, of seven or eight inches wide and the length of the frame, before him; this board is so fixed to the frame that it may be moved, over again, and near the spindle; he has the cotton in a basket near him, and with his left hand spreads it on this board along the spindles which by their turning draw the cotton through them being wide enough to admit the cotton, but too near to permit the seed to go through, which being thus forced to leave the cotton in which it was contained, and by its rough coat entangled; falls on the ground between the workmans legs while the cotton drawn through falls on the other side into an open bag suspended for that purpose under the spindles.

The French in Florida have much improved this machine by a large wheel, which turns two of these mills at once, and with so much velocity as by means of a boy, who turns it, to employ two negroes at hard labour to shovel the seed from under the mill: One of these machines I saw at Mr. *Krebs* at *Pasca Oocooloo*, but as it was partly taken down, he claiming the invention was very cautious in answering my questions, I cannot pretend to describe it accurately; I am informed that one of those improving mills will deliver seventy or eighty pounds of clean cotton *per diem*.

The packing is done in large canvas bags, which must be wetted as the cotton is put in, that it may not hang to the cloth and may slide better down; the bag is suspended between two trees, posts or beams and a negro with his feet stamps it down; these bags are made to contain from three hundred and fifty to four hundred weight; with about twenty slaves moderately working a very large piece of poor ground might be finely improved so as to yield to its owner a fine annual income by means of a staple which is much in demand in England, and here is raised by no means inferior in whiteness and fineness as well as length of fibres to that of the *Levant*.

XXII. The *Mulberry*, deserves our next notice for a two fold cloathing it provides us with.

Among my botanical articles the reader will find the description of one of this class, which I have all the reason in the world to believe to be the *Morus papyrifera* and which I have on *page 20 of this volume* distinguished by the name of *Morus foliis palmatis, cortice filamentosa, fructu nigro, radice tinctoria.*

This tree is found in abundance in the North Western parts of Florida: The chactaws put its inner bark in hot water along with a quantity of ashes and obtain filaments, with which they weave a kind of cloth not unlike a coarse hempen cloth; I would propose to boil the bark in a strong alkaline *Lixivium*, by which means I make no doubt but a very fine and durable thread of the nature of cotton, flax or hemp might be obtained; the root of this same tree likewise yields an excellent yellow dye: But I shall here treat of the article, which is most commonly known to be produced by means of the mulberry tree, this is silk: A very short time about six weeks in the year will suffice for all the labour requisite to acquire this valuable article, and that labour is so light as only need children to attend it.

The gathering of the leaves being the most laborious part of the work I would advise the sowing the seed as it were at broad cast, so that it may spring up in form of wide hedges of about ten feet breadth leaving a lane of two feet between each pair, by this treatment the leaves may be gathered by means of a pair of sheers or if the hedges are narrower the hand may do it, without the disagreable necessity of climbing trees, which is always more or less attended with some danger, and as this is a female business, with indecency.

I am convinced mulberry bushes will grow thus and yield abundance of leaves and therefore this method is eligible before groves or orchards, which take up much room and have a dirty effect during the fruit season.

All the species of mulberry trees grow kindly in Florida and some people pretend the white kind to be best, but on my strictest enquiries I could not find in what manner this affects the worms, but I would have the silk planter be very cautious if he has one kind in his nursery, strictly to banish the other, because this change of leaves is certainly the reason of some of the diseases attending the worms: The remainder of the silk culture is no more than to keep the worms well fed, and the apartment where they are kept thoroughly clean;

when they begin to acquire a certain transparency, the period of their spinning or resolving themselves into a *Chrysalis* is at hand; then it is necesary to put up bundles of some slight thin twigs between the shelves. The *wild* or *dogs fennel* affords a ready and proper material for it; here the worms will naturally enough mount upon and pitch on a place where to metamorphose themselves into a cocoon; in Georgia we have a filature, likewise at Purysburg; but as there are none in Florida I will subjoin the following account of its preparation for the manufactury.

1st. The cocoons are to be put into an oven just hot enough to deprive the *Chrysalis* it involves of life, without hurting, the fibres of the cocoon:

A heat something below *Fahrenheits* scale for boiling water will effect this: without this precaution the insect eats its way out and destroys the thread of silk.

2dly. It must then be put into a copper with water just on the boil and kept so, this will discharge the glutinous matter from the cocoon and discover the end of the clew; then taking several of these ends together they are gently reeled off, and afterwards spun and prepared for the loom.

This process is hurtful to the elasticity and strength of the silk, though it does not deprive it of its gloss: Therefore if we could attain the knowledge how the raw silk is managed in the *Levant*, it would be the most eligible way; all we know about this is, that it is performed without hot water: this is called raw silk and comes in bales to England and other manufacturing countries.

The refuse cocoons either damaged by the insect or other ways, are carded in Europe, and are then improperly stiled raw silk; this should not be confounded with the above named from the *Levant*, being by no means equal to it. After the silk is reeled off, we find some irregular coarser kind on the inner division of the cocoon; damaged cocoons are mixed with this, as also the inner division next over the *Chrysalis*, after being steeped in warm water to dissipate its gelatinous parts; this mixture is carded and called *floretting*.

All these carded silks loose their lustre by that process.

Of this commodity there is at present imported into Britain

*From Spain and Italy to the value of ⎱ £. 1,500,000 ⎰

---

*Thoughts on the times, and the silk manufactory 8vo. page 7, anno 1765.

*From Turkey, the East Indies, and China   400,000

£. Sterling 1,900,000

This account shews sufficiently how beneficial this business would be to America, would we by our industry in a very light labor try to oblige England to let this money circulate here instead of among Spaniards, Italians, Chinese and Turks; which is by no means impossible, since the bounteous hand of nature has furnished us with the requisite plant among the number of those that are indigenous in Florida.

XXIII. The *Carthamus Tinctorius* or safflower is likewise found among the indigenous plants in the western part of Florida; this useful dying weed requires but little cultivation, and well deserves our attention in that country.

XXIV. *Hemp* and *flax* (according to a very ingenious performance†) Britain imports from the Baltic annually to the value of 500,000. £. sterling; another great encouragement for Florida; neither Carolina nor Georgia have any lands comparable to our fine lands on the Mississippi, and yet they have already exported considerable quantities of hemp, and thus set Florida an example well worthy of imitation; the lands are so rich on the Mississippi, that neither of these two impoverishing plants will exhaust them, therefore let us apply ourselves to this cultivation, which is so universally known as not to need description; thoroughly pulverizing the earth, and not sowing it too thick are almost the only things to be attended to in its cultivation, and the proper criterion of rotting the *ligneous* parts of the plant, so that they may be easily separated in the brake, is the only one of moment in preparing it for embarkation.

Add to this, that ere long we shall have extensive settlements producing immense quantities of materials for exportation on and near the banks of that almost unbounded interior ocean the Mississippi, for three thousand miles up it; not to mention the products of the river *Ohio*, the *Shawanese, Ouabache, Hogoheegee, Yasoo, Missouri, St. Peter, St. Francis,* and the red and black rivers with many others of inferior note, all emptying themselves in it, where there is so much

*Political essays concerning the present state of the British empire, 4to, page 191, London 1772.
†Museum Rusticum, vol. 1, page 457.

room for the increase of people; which always proceeds in proportion as there is more space for them to sit down in, this is beyond reply verified by so amazing a rapidity of increase as America has experienced within these twenty years, being no less than in triplicate proportion. Now it is evident, that to carry off the produce of this vast tract, it will be necessary to build ships in every part of it which together with their bulky commodities must be sold abroad, as very few small craft will be sufficient to bring up the trifling returns the inhabitants of this happy country may stand in need of; this being the case let us consider that timber, iron, lead, &c. are found up this river, but without rigging and sails they cannot constitute a ship; likewise we must recollect, that rigging and sails are bulky articles and would cost much for carrying up so immense a distance. Think not reader, that these are chimerical ideas, by no means; every part above named has already in some degree experienced more or less the effects of the industrious ax and hoe in the hands of the Herculean sons of America.

There is a very strong kind of fibre called Indian hemp in Florida; I would recommend an inquiry into what it is; the savages use it and I am persuaded we should find it worth improving.

The use of flax is too well known and its necessity so evident, that a description or recommendation of its culture and preparation would be superfluous.

The North American *Annona*, the *Lime*, and *Mahoe* tree; all indigenous in Florida, yield each a serviceable bark of great use if properly manufactured.

XXV. *Tobacco* is a source of great riches in this country; the French have proved that this plant may be produced in great quantities and of the best quality; this may be made an article of great emolument especially as that trade is excessively on the decline in Virginia and Maryland, but if we cannot keep it clear from the multiplicity of incumbrances it is saddled with in those colonies, such as a trade by factors and many etcaeterae which makes its value to the planter very inconsiderable, I would rather cultivate none; it is a rank luxuriant, impoverishing vegetable, requiring a deep rich soil such as the Mississippi alone affords here; therefore it will undoubtedly thrive; its culture is almost identically the same as that of the *Zea*, but I would advise the planting it further apart, such as six feet at least; the only material thing this culture differs in, is because the seed being exceedingly diminutive in size it must be sown in beds of rich new ground, which has been well manured, with the ashes of the brush cut off from it; from hence it is transplanted in regular rows at the above named distances: In about three weeks time it will be advanced to the growth of a foot: Now the first hoeing takes place, and suckers and worms begin to appear; from both these the plants must be carefully cleared.

The plants are at this period out of danger of being scratched out of the ground by a large flock of turkies which may now be turned into the field, who will not touch the plants but carefully look for the worms that infect them, of which those birds are very fond; and thus they will save a great deal of labour, but the suckers must be attended to by human labour, which is also required to keep the ground clear from weeds; when the plant attains the heighth of eighteen inches it must be deprived of its head and stripped of its lowermost leaves, reserving a few with their heads on for seed: In about eleven weeks after they are transplanted the plants cease growing and the leaves assume a variegated colour of different green shades; this is an indication of their maturity; they are then cut and laid in heaps to sweat for about sixteen hours; next day they are carried to a shed, so constructed as to admit a free circulation of air through every part thereof, but well covered against rain; this is the practice in Virginia; in some counties they keep all the light out except where the rays cannot immediately reflect on the tobacco; in this shed they are hung with their tops downwards for about six weeks; then taking advantage of a day when the atmosphere is pretty much loaden with moisture, they are laid on a kind of floor of poles, and left to sweat for about ten or twelve days; now the leaves are stripped off and the tops kept separate from the middle and the lower leaves; the uppermost leaves being the best tobacco and the lower the worst; they are then made up into very close packed bundles called in Florida *Carrots*, which are held together by a string of bark; there are at present but two sorts produced viz *Nanquitoche* and *Pointe coupée*, the first infinitely superior to the second.

XXVI. Pitch, tar and turpentine being the produce of vegetables will not improperly come in here; the process of making the common tar is by splitting the heart of the pitch pine, fallen down (with which the ground is covered in the eastern and southern parts of Florida) into small sticks of the length of about three feet, and arrange them into a kiln of a circular form; but as this is a process hardly to be learned by theory, I will recommend it to the inquisitive reader to inform himself at a tar kiln great or small, which are a very common occurrence throughout this continent.

Sometimes this kind of tar is made of pine knots without splitting.

The green tar has not yet been made in Florida and until I gave some account of the process to some of the planters there, it was intirely unknown in the country.

This is made from trees that have been tapped for turpentine, and continued running for about three years, by chopping the bark off by degrees to twelve or fifteen feet high; they are then cut down and split into small sticks; and the tar is made by the same operation as the common sort.

Notwithstanding the great quantities of pitch pine in Florida, turpentine has not yet been made there; even in February 1773 when I was at *Orleans*, there being a necessity for some turpentine on board of the vessel (in which I came from thence) the Captain was obliged to pay sixteen dollars for a half cask of it, and I have been informed, that even then it was sold in that town for medicinal uses at a great price by the pint and quart.

The process is very simple; a hole is cut in the tree on the side most exposed to the solar rays (in large trees two of these may take place) these holes are cut slanting downwards leaving the bottom concave and large enough to introduce a calebash to sip out the gum, which deposits its self in the bottom of the hole; this hole is called a box, and the turpentine is dipped out of it and put in casks: Florida, especially the Eastern and Southern parts, has the advantage for this business of the more Northern provinces by not being tied down to any particular season, the climate admitting of this work the year round.

Pitch is in this country generally made by burning the tar in the clay hole into which it runs from the kiln but I would recommend the use of a still for this purpose, which is agreeable to the practice in Sweden and Germany; by this means a very valuable essential oil is saved which arises in the boiling; this oil is called *Oleum pini* or *Oleum taedae*.

It is of great use with painters, varnishers &c. on account of its drying quality, it soon becomes thick of a consistence like balsam:

Along with this oil comes over a watery liquor, which the workmen injudiciously throw away; it is a good acid spirit, capable of being applied to sundry useful purposes: Neuman says he knew a person in France, who had saved by it several thousand dollars.*

Resin commonly called Rosin is best made by distillation also, a considerable quantity of oil would be obtained, and the expence of a still would be amply repaid by these savings.

As pitch and tar are often used to pay roofs of houses &c. and the addition of red lead &c. made use of, which has proved ineffectual, and as the same observation takes place with regard to ships bottoms, were water (as does air in the former case) corrodes the pitch, I think it not improper to insert the following note of the ingenious Dr. Lewis on Neuman's chymistry: "An anonymous correspondent of the Swedish academy observes, that the tarred or pitched boards or shingles, with which houses in many places are covered are soon damaged by the heat and moisture, the sun's heat melting off the tar so that the wood remains bare. Some have endevoured to prevent this, by laying on the tar late in the year, that the winter's cold might fix it; an expedient of some use, but very far from being effectual, for however the tar may be hardned by the winter, the return of summer soon makes it soft again. Others have mixed smith's cinders with the tar, which instead of mending the matter makes it worse, all ponderous substances do the same. He says, he has never found any addition comparable to coal dust (that of charcoal he seems to mean) which is to be stirred into tar made hot in such a quantity as to make it thick: The mixture is to be laid on with wooden trowels in a hot day: tar thus prepared, he says, is fixed, never runs, binds and hardens surprisingly from heat and moisture and shines better than oil varnish."

Pitch burnt in a close furnace, so that no smoak escapes causes a soot to arise, which is the lamp black so much in use.

The last quoted gentleman in the same page* favours us with the following annotations: As I think other words than his own, cannot be so descriptive of any process whatever, I hope my reader will excuse these quotations:

"What is called lamp black (originally perhaps the soot collected from lamps) is obtained in different parts of Germany, Sweden, &c.

---
*Neuman's chymistry 4to. page 88, London 1759.

not from pure resin or pitch, but from the dregs and pieces of bark of the trees separated in their preparation, for making common resin, the impure juice collected from incisions in pine and fir-trees is boiled down with a little water and strained whilst hot through a sack, on cooling the resin congeals upon the surface of the water, and is then packed up in barrels; it is distinguished according to its color, into white, yellow and brown. The dross left on straining is burnt for lampblack in a low oven, from which the smoke is conveyed by a long passage into a square chamber, having an aperture in the top, upon which a large sack is fastened: the soot concretes partly in the sack, which is occasionally removed, and partly in the chamber and canal, from which it is swept out."

XXVII. *Gum elemi* is a resin fit for many medicinal uses, and the product of a species of *Pistachia*, very common in the southern and southeastern parts of *Florida*; the wood in itself is not valuable, but as the gum is one of those articles, that are generally brought to England in a sophisticated state: this tree is worthy our attention, and may prove a serviceable article to the country; but this is not the only use of this tree, it is very useful for cattle either in dry seasons in the southern part of the country, or for such as are kept up at the farm yard, they are very fond of the leaves of this tree, which are a wholesome food, increasing milk, and keeping cattle in a thriving condition.

XXVIII. A species of tree greatly resembling the poplar is found in *West-Florida*, which produces, if not the *Tacamahaca*, at least something of the resinous kind so extremely like it, that it is not distinguishable from it; I am told, that the resin is obtained by digesting the buds, which appear in the spring.

XXIX. The *Liquidambar*, or maple leaved *Storax* is also found in abundance in both Floridas; from this the *Storax* is produced, by boiling the branches; this valuable gum may be worth attention, as a great deal might be gathered in these provinces.

XXX. *Cassia Cinnamomia*, or *Cortex Winteranus*, is an article found in abundance in the southern parts of the *Peninsula*; the people from Providence know it by the name of *Cinnamon*, and carry a good deal of it to market; it is a medical plant of good use, and worth our notice.

XXXI. *Balaustians*, or the double flowering *Pomegranate* comes to a good perfection here, and as they are a profitable article they deserve our attention.

XXXII. *China Root* is found in both provinces, but the prickly plant, commonly known by that name, does not yield the genuine kind; the true one, which is not thorny, is also found in abundance; and as in *India* this is paid for at the rate of from 6 to 16 shillings per hundred pounds, to send to Europe, it undoubtedly is an article worth exporting from America.

XXXIII. *Ipecacuanha* is found in almost every spot of oak land in this country, it may deserve to be manufactured from the spontaneous root, if not to be cultivated.

XXXIV. *Jalap*, an article of consequence in trade, not only on account of its use in physic, but likewise for its universal use in the fermentation of liquors: Europe has hitherto been obliged to import this article from Mexico, in which kingdom is a place called *Xaleppo*, or *Haleppo*, of which this drug bears the name; the only place where it was supposed to grow: we have hitherto been at a loss to know the genus it belongs to, and many roots of purgative quality have been supposed to be it, and were substituted in its room: the late *Doctor Houston* introduced it from *Mexico* into *Jamaica*; but while he was gone to *England* the man whom he left in care of it, suffered hogs to destroy it; however, this gentleman brought a pencil drawing of it to Europe, but as this did not shew the colour, and the seed has been sown in the botanical garden, at Chelsea, without success, what it was remained still a secret, until I accidentally found it growing wild near *Pensacola*; being led to think, that a certain tuberous root made use of by the savages as a purgative might be it; I dried some slices of it, and found it so nearly agree with that drug in appearance, that it caused me to examine all the *convolvuli* I could find in this country, because I was informed that to that genus the plant belonged: I succeeded, and samples, which I sent to divers parts of Europe and America, have proved to be it, and of a good quality; this plant is pretty plentiful in some spots on the highest and driest lands, and I suppose its cultivation must be somewhat analogous to that of carrots.

XXXV. The above researches among the *convolvuli*, have made me acquainted with another of that *genus*, which seems to answer to the *Scammony* plant; I am not yet able with certainty to determine whether it really is the Scammony or not, but have great reason to believe it is.

XXXVI. Seeds of the true *Rhubarb** having been lately introduced into *America*, and the deep soil of the *Mississippi*, being very proper for its cultivation, I would by all means recommend trials to be made for raising that valuable root, which I think will not fail to answer our expectation.

XXXVII. A shrub resembling in its fructification the *Starry Anise* is found in *West-Florida*, but it totally wants the aromatic taste of the seed; perhaps cultivation may bring it to the same state of perfection with the oriental.

XXXVIII. *Silk Grass* grows on the most barren sand hills of *Florida* (called black Jack ridges) if it does not deserve a cultivation on account of its fibres; yet the root having been found by experience to wash woollen the cleanest and whitest of any thing yet known. *Quere*, would it not be useful for the woollen manufactory?

XXXIX. *Arnotto*, a useful dye is introduced in the colony; I have seen some of the plants vigorous and in good health, in Mr. *Wegg's* garden; should it succeed, its cultivation is by no means to be neglected, however, if it fails in *West-Florida*, the southern parts of *East-Florida* will certainly produce it.

XL. My knowledge of *Mosses* being hitherto very confined, I shall not speak with certainty about the *Argal (Lichen Rocella)* but having seen some *mosses* on the rocky islands, and part of the *Peninsula* greatly resembling the *Argal*, gathered on the island of *Orchilla*, I verily believe this valuable article to be there, if not it might easily be introduced from *Orchilla*; the face of this part of

---

*In *Bossu's* travels by *Foster*, vol. 1, page 353, London 1771, mention is made both of *Jalap* and *Rhubarb*, as indigenous in *West-Florida;* my strictest inquiry on the subject near the banks of *Mississippi* have been in vain to find *Rhubarb*, or even any thing the *French* call so, and as for *Jalap* I have all the reason in the world to believe, that this superficial writer never was in those parts of *West-Florida* where it is found, as he wrote his book merely to blacken the officers of a different opinion from himself, in regard to their conduct in admitting *English* vessels in time of need, into *New Orleans*, and being an inveterate enemy of M. *De Kerlerec* in particular, cloaked it under the specious title has has done, and therefore I believe, that he received all he says, concerning natural history, from hear say only, and thus has been led to mistake a kind of *Belle de Nuit* with a purgative root, which is pretty common on *Mississippi*, for the true *Jalap.*

Mr. *Foster* in his note on this passage has copied the same mistake, when he says it is the *Mirabilis.*

the country exactly resembling that island, and its climate being nearly the same.

XLI. *Cochineal.* This valuable insect is found in pretty large quantities in the *Floridas*, especially in the eastern province, of the kind, which is called *Sylvester*, on a species of *Cactus*, or *Opuntia;* could the true *Cochineal Cactus* be introduced into *East-Florida*, I make no doubt of its succeeding and the vicinity of the *Cochineal* countries makes this not at all improbable; the people from the *Musketo Shore*, or *Bay of Honduras* might be serviceable in obtaining it.

XLII. *Sumach* greatly necessary in dying and tanning, which is much used for preparing Turkey leather; several kinds grow in the southern parts of *America*, and therefore it is worth our while to enquire into the matter, to know which kind is used for this purpose. This plant is also known in medicine. The large kind is used to make vinegar with, and I am told by the *French* people, that a piece of the wood put into a cask of weak or faded vinegar, or even water, will produce an intensely sharp vinegar.

As a hint to travellers, in the southern parts of *America*, where the distance between the settlements often obliges us to carry our provisions with us, I will assure them, that the fruit of this kind, steeped a very short time in water, communicates to it a very agreeable acid flavor, which will render it very fit to make punch; which agreeable liquor proves a great refreshment in a hot day, in the woods.

Another kind possesses a noxious quality; this grows in low grounds; beware therefore of making spits of this to roast meat on, and take it for a general rule not to employ any wood (unless you are well acquainted with it) growing in low grounds for that purpose, as almost all the noxious plants, of this country, are found in such situations; in high grounds you may indiscriminately use any wood (which has no bad taste) for that purpose.

XLIII. If the acorns of the *Quercus Suber* could be introduced here, I make no doubt, in this part of the world (where above twenty kinds of oak are indigenous) the cork manufacture might be profitably carried on.

Since I am on the subject of oaks, I beg leave to mention, that in *New-England* there are works for making an extract of oak bark, which yields a considerable advantage by sending it to *Europe* for the use of the tanners.

I have seen also a kind of dwarf oak producing galls; *quere*, could they not be made use of?

XLIV. *Madder, Rubia Tinctorum.* This root is one of the most useful ingredients in dying wooll and stuffs red, as also cotton of an agreeable bloom color, and consequently much used in *England* for the different manufactures; but incomparably the greatest quantity used is imported from abroad, to the amount of large sums.

It is likewise said, that *Madder* is an excellent food for cattle, that it increases milk, and causes the butter to have a most agreeable color and flavor; I know it to be sometimes cut for hay and we are told, that it makes an excellent fodder.

This plant does undoubtedly deserve our attention in Florida, on the above accounts, especially as the many trials to grow it to advantage in *England*, seem for the most part unsuccessful. Many different kinds of madder have been tried for this purpose, but none have yet proved of real use except the *Rubia Tinctorum Sativa* of *C. Bauhine*, which is the sort cultivated in *Zealand*, and some parts of *Austrian Flanders*.

If it be objected to the culture of Madder in *Florida*, that these are in a very different climate from the southern part of *British America*, let it be remembered that in the *Levant* it is cultivated with success, and that what comes from thence is a more valuable dye.

The ground in which Madder thrives best seems to be a deep black mould, in something of a low situation, which should not have a clay foundation, but sand or gravel; the land in *Zeeland* is, and that on the river *Amite*, seems to be in general of this kind.

It is cultivated in *Zeeland* by offsets, or shoots, which they take from an old plantation, and replant immediately in rows, about eighteen inches apart; the young plants have each a distance of four inches allowed them, and the ground is divided into beds of twelve feet wide, leaving a ditch of about twenty inches between them; this is done in the beginning of May, and great care is taken that no offset is planted, without it be furnished with fibers; as it is thought that for want of fibres they would miscarry, which they often do even in the most favourable seasons. The greatest labour, I think, I have seen the people at in *Holland*, in regard to this culture, is the covering the stalk, when they attain the height of about sixteen inches, leaving only the tops bare, in order to promote the multiplication of roots, which is the part of the plant manufactured and sold; when this covering is performed, there remains only the attention of weeding, which ought to be done often; the root is generally taken up the second year, but I think I have heard it said, as well as read it, that three summers are necessary for this crop to come to full maturity; the roots, which are said to yield the most and best dye, have been taken up, when they had obtained about three tenths of an inch diameter in size; it is thought that when they grow too large, they yield a dye more inclining to yellow than red.

The lateral fibrous vermicular roots, are said to yield a superior dye, but not to pay for the expense necessary for gathering them.

From this general sketch of the Madder culture, such as it is in *Holland* and *Zeeland*, my readers may see, that it is not so expensive an affair as it is generally deemed to be, but like all other things the cultivation of this plant may be carried on at too costly a rate, and it likewise may be attempted in too penurious a way: I have endeavoured to make my writing intelligible to every capacity, and therefore hope, that every one of my readers may be led into the true idea of this culture, to make it answer the purpose in *Florida*.

This cultivation by sets or shoots being practised in countries where the seed does not at all, or very difficultly come to maturity, I think the seed ought to be introduced into *Florida*, or endeavours made to obtain it, from plants to be carried or transplanted there, which, if productive of seed, it ought to be sown in drills, like rice; which I would think the most eligible in the moderate climates of those provinces. I believe this plant to be a great impoverisher of the soil, for in *Zeeland* they always allow some years between every two crops in the same spot.

As it has lately been said, that there was no necessity for drying Madder, and that in using it green, there is even in the evaporation of dying matter a saving of one half, besides the greater saving of the expenses of a kiln, a mill, or drying house, &c. I must inform my reader, that he will find all this true, but then it will be necessary for him to transport the dying houses from *Europe* to our Madder fields, and not the Madder to the dyers, in order to enjoy the profits of all this great and oeconomical frugality; for perhaps there is not a plant on earth so soon inclining to fermentation and putrefaction, which is occasioned by its succulency; yet for the planters present family-use it is certainly fit to use green; as soon as the roots have become spotted, or black, or lost a strong scent (similar to that of liquorice) they are utterly unfit for any use; I shall therefore make a few remarks, necessary to be known for the drying process in *Florida*. A hot, sun-shiny day may be used to advantage, to dry the roots partially; but if the weather be not favourable, when the roots are taken up, they must be spread within doors on a floor, taking care to spread them thinly, and stir them often; but this will never absolutely preserve them from changing, much less make them fit for transportation to any distance; if the crop be small, a baker's oven may suffice, but beware of raising the heat above 180 degrees of *Farenheit's* thermometer in the place where the roots are put, which should be over the oven; but for larger crops kilns, similar to malt-kilns are necessary; take care to make them roomy, keep an equal and moderate heat, and by all means prevent any the least access of smoke to the roots; for which reason I would advise large ovens, such as the biscuit bakers in *Holland* use, as preferrable to every other method; a building may be so contrived as to contain 13 ovens, viz. four on each side, three at one end, and two at the end where the door is, with one general brick floor over all; one or two windows may be so contrived as to give access to a sufficient light; let us suppose the oven ten feet long, by eight wide, and allow two feet for each partition, this will make an oblong apartment of forty-two feet by thirty-two in the clear below, and on the upper floor fifty-two, by forty-two, room enough for any crop; provide good brick funnels to your chimneys, and there can be no danger of fire, the rest of the building may be of timber; in this process Madder will loose five sixths of its weight.

    When the roots are sufficiently dried, they must be pounded in wooden mortars; for this purpose a mill constructed exactly like the old fashioned rice mills is very proper, only varying in the shape of the lower end of the pistil, or beetle; for in the rice mills their lower end is in form of an inverted cone; but here the lower end ought not only to terminate in a square, but the but-end ought to be cut into small squares, so as to render the pestle toothed; for this reason also the mortar ought to be of a different form from the rice mortar, which last is likewise an inverted cone, or shaped like the side of the top of a funnel, whereas this ought to be in form of a hollow globe, which has a neck like a decanter or bottle, in which neck the pestle ought nearly to fit. To empty the mortars and supply them with fresh roots, is a necessary occupation during the pounding. The roots ought to be cleared of their outer bark.

It will then be fit for packing into casks and exporting.

I remember to have heard it said in *Holland*, that poor people, in order not to be obliged to sell their small crops to the manufacturer, at his own price, preserve the roots from fermentation, by burying them between layers of earth in the ground, and that by this means it may be preserved for any term of time, without perceptible alteration.

XLV. *Coffee* being an article to all appearance fit to be raised in the southern parts of the peninsula, and some of the islands, I shall give it a place here. This is an object worth our notice, as its consumption is great already, and still increasing; there is little or none produced in the *English West-India* islands, in comparison of that which the *Dutch* and *French* colonies produce.

Its culture throughout all the settlements where it is raised, is now by the young shoots obtained from the larger trees; but originally it was introduced by seed, which being soaked for about twenty-four hours, is then planted in tubs, pots, or beds, at about three inches distance, covered lightly with earth, and carefully watered when no rain happens. Usually in about 14 days time the plant appears; when the young plants have attained the height of eight or ten inches, a rainy day is watched to transplant them into a walk, as it is there called, which answers to our orchards; the ground is here carefully cleared of all manner of roots and plants, and turned up at least a spit deep. About twelve feet is the distance which ought to be left between every two plants.

The growth of coffee is quick, provided the ground be kept clean, but perhaps no plant is sooner hurt than this by too luxuriant a growth of weeds or plants, round or near it. The second year *Eddos*,† or *Taniers,* or even patatoes, may be planted among them, which will be a means of raising provisions by the same labour, that is necessary to keep the ground clean.

This plant bears fruit sufficient to defray the yearly expences at the end of the third year; its produce will then increase until the seventh year, and after this it will continue to bear in a degree nearly equal, until about the fortieth year of its age, when it begins to decay.

---

†*Eddo* or *Tanier* is a species of esculent *Arum,* well known in *East Florida,* and is good food for negroes.

If any of the young plants should fail, they ought immediately to be replaced by others.

In the *Dutch* colonies, when a coffee walk decays, they root the trees out, and let the ground lay fallow for ten years, or upwards, during which time it affords pasture for cattle, and afterwards it is turned into a cocoa walk, or cotton plantation.

The ordinary height of this tree is from twelve to sixteen feet; in the *Dutch* colonies they are lopped, to reduce them to a kind of *Espaliers,* for the easier gathering of the fruit.

When the coffee has attained to ripeness, it is carried to drying sheds, which are constructed in the *Dutch* colonies on the following plan:

The *Area* of the ground generally applied to this use is seventy feet by thirty; a brick foundation of four, five or six feet high, is first laid to raise the building from the ground; on this the building is placed of timber, being of two stories; the upper floor is about twelve feet above the lower; in each side of the building are from twelve to sixteen large windows likewise two at each end, on each side of large doors, all this is necessary to give a free access to the air, to prevent the coffee from heating or shooting. In the lower part of the building a kind of drawers, of about six feet square are so contrived as to be drawn without side the building, where they rest on wooden rollers or blocks, these drawers receive the coffee after the drying floor, and in fine weather are drawn out, but pushed back the moment it begins to rain; by this contrivance a large quantity is, as it were, instantly sheltered from ruin, no other invention can be so expeditious.

The building must be furnished with two pieces of square timber, of the length of twenty-five or thirty feet, and about eighteen, or twenty-four inches thick, made of hard wood; in these a row of mortars is sunk, to beat the coffee in, pestles or beatles for the same, fanning mills to clean the coffee, shovels for turning it often while it lies on the upper floor; a competent number of baskets, of different sizes, and a pair of scales with weights. Before this building there are generally one or two platforms, from forty to fifty feet square, called drying floors, intended to take all benefit of the fine drying weather during the coffee harvest. Adjoining to the building is generally a smaller one, containing a cooper's shop and a mill, called a breaking mill, through which the new gathered coffee passes, to deprive the grain of its pulp, or red outer skin; after coming from this mill it is soaked during one night in water, and next day spred upon the drying floors, where it remains till the air and wind have sufficiently dried it; if rain should happen it is quickly gathered into heaps, and covered with sear cloth.

The coffee, being thus dried, is put into the drawers, where it is left till thoroughly dry; from hence it is carried into the loft (being now only surrounded with a thin, semi-transparent husk over each pair of seeds) where it is left till the whole crop is gathered; the harvest lasting often two months; while it remains in the loft it must be daily turned, to prevent its heating, and in good weather all the air possible must be admitted; after harvest it is again returned into the drawers, and left there for three or four days, in order to become totally dry, it is then pounded or beat in the above mortars by hand, to deprive it of the thin, inner husk, which involves every pair of seeds; after this it is fanned, and when fanned the broken grains are separated from the whole, which last are put in bales, or casks, for the market, with all this seeming trouble a coffee walk is easier attended than a sugar plantation, and is said to be full as profitable.

XLVI. *Cacao* is a plant which I make no doubt would succeed in some of the lands found in the southern division of *East-Florida*, but as it is tender, and requires a deal of attention I shall only hint at it, as an object worth trying, and we ought to do this out of regard for our lives and constitutions; chocalate being become so common an aliment, and the vilainous adulterations of it in the northern colonies, make it an enemy to the stomach, whereas if good and genuine it proves a very agreeable, nourishing and balsamic food.

XLVII. *Tea*, a despicable weed, and of late attempted to be made a dirty conduit, to lead a stream of oppressions into these happy regions, one of the greatest causes of the poverty, which seems for some years past to have preyed on the vitals of *Britain*, would not have deserved my attention, had it not so universally become a necessary of life; and were not most people so infatuated as more and more to establish this vile article of luxury in *America*; our gold and silver for this dirty return is sent to *Europe*, from whence, being joined by more from the mother-country, it finds its way to the *Chinese*, who, no doubt, find sport in this instance of superior wisdom of the *Europeans*. These considerations, joined to the additional evil of its being a monopoly of the worst kind, and the frauds of mixing it with leaves of other plants, ought to rouse us here, to introduce the plant (which is of late become pretty common in *Europe*) into these provinces, where the same climate reigns as in *China*, and where (no doubt) the same soil is to be found; by this means we may trample under foot this yoke of oppression, which has so long pressed the mother country, and begins to gall us very sore; and will the *Europeans* (according to an unaccountable custom of despising all our western produce, when compared to oriental ones) avoid drinking *American* tea? Be not ye so infatuated, ye sons of *America*, as not to drink of your own growth! Learn to save your money at home! I cannot think this advice contrary to the interest of *Britain*, for whatever is beneficial to the colonies, will in the end be at least equally so to the mother country.

In hopes of some well minded planter taking this into consideration, I have thus hinted at it, and will write what I can, from hearsay, of its culture. First then let me caution the attempter against imposition; be sure that you get the seeds in good order, and in their capsules, for the friend you employ may be deceived more in the plants than the seeds, there being a plant called by *Linnaeus*, *Camellia*, and by *Kaempfer*, *Tsubakki*, so exactly like to Tea in its leaves, as not to be distinguished therefrom; this has frequently been sold to the *Europeans* for Tea, and is thus introduced into *Europe*. If the seed cannot be procured in *Europe*, nor the real plant from a trusty friend, apply to some well-meaning Captain of an East Indiaman, no matter of what nation, he can get the seed in *China*, and it may be brought to *America*, without any danger of decay, in a manner which I shall hereafter mention for the preservation of seeds in general. After the tea seed procured some may be sown on the passage from *England* here. Will any reader say this is building castles in the air? let him remember, that seventy years ago rice was an utter stranger in *America*, till some such good captain, who had been in the east made a present of some seeds to a person in *Carolina*, who planted it; both these men are seemingly forgot, but how much more did they deserve statues, than many others, who have been in an unnatural manner, often for crimes, idolized, and as it were eternized, by pompous effigies, and lying monuments!

Even in their native country the seeds are very liable to miscarriage, therefore they plant ten or twelve in each hole, in rows throughout the field; *Kaempfer*, who had the best opportunity of information, tells us, that the plants are left to grow till seven or eight years before the leaves are gathered; but he should have told us, whether this was necessary to the perfection of the leaves, or only done on account of the quantity obtainable, perhaps to strip them younger would destroy the plants: I think he also says, that after ten years the plant is cut down, and young shoots spring up, or the plantation is entirely removed.

In the gathering, it is said, much time is taken up, each person culling not above two or three pounds per day, it being done leaf by leaf; during the time a sheet is spread over the bush, in form of a canopy, to prevent the leaves drying hastily, in the sun-shine, which would exhale the flavour; it is said also, that they have three crops, viz. in the end of *February*, this is the best; then in the end of *March*, and beginning of *April*, and the last and worst is in the end of *April* and beginning of *May;* what remains on the bush till the end of *May* is said to be unfit for curing, and consequently for exporting; the observation on tobacco is also said to hold good in regard to tea, I mean, that the upper, middle, and lower leaves are different in goodness, and consequently three different kinds.

The curing is done on iron plates, over a moderate artificial fire, and the leaves are constantly rolled by the hand. *Neuman* thinks the dark colour, and the rose flavour of Bohea tea, to be artificial; he seems to say that *Kaempfer* and Dr. *Cunningham* (when they wrote that Bohea tea was the first gathering) meant it to be a crop of leaves collected before the plant comes to maturity, which is perhaps the case in the provinces of *China* that are too far north, to bring it to perfection: He also mentions the process related by one Meister, to wit, that the leaves are put into a hot kettle, just emptied of boiling water, and that in this they are kept close, till they are cold, afterwards committed to the hot plates this produces Bohea tea; he adds, that this at least is certain, that by a similar process good Bohea tea may be made among ourselves; a specimen of which he himself exhibited to a numerous audience. He further says, some dealers in tea, in *England* are not ignorant that certain *European* leaves, particularly those of the *Dog Rose*, and *Cherry tree* may be so coloured as to pass for good Bohea. The leaves of the red whortle, or cranberry, and rose bush seem now to be most in esteem among these sophisticating gentlemen.

The above author relates, that the *Chinese* are extremely curious in every branch of this manufacture, they gather the leaves with thin gloves, and the workmen employed in cutting it are restrained for a fortnight before from flatulent food, or whatever else may communicate an ill flavour.

There is a later writer on this subject, I think Dr. *I. C. Letsom*, who is perhaps better informed, but I have not been able to consult him: be all this as it may, let us introduce the plant, and we will discover its culture and curation.

XLVIII. Such numbers of *Lauri* being found indigenous in America, it may not be amiss to hint at the *Pimento*, or *Jamaica* pepper, as very proper to be introduced into the southern parts of *East-Florida*, where, I make no doubt, it will prove a profitable article.

XLIX. *Mangroves* and *Salsola: Pot, Pearl* and *Barilla ashes*, can nowhere be made to more profit than in *East-Florida;* the several *genera* of trees known by the name of *Mangroves*, are in so great an abundance, and so replete with salts, that they will undoubtedly yield a much greater proportion than any wood yet used for that purpose; I have made several very good samples on the coast; these trees have never yet been properly classed, I intend to do it in the course of this work: the *Kali* for *Barilla* is perhaps nowhere so abundant as in that province.

L. *Bees* are in great numbers in the eastern province, but found their way as far westward as *Pensacola* only about the year 1772, they not being original natives of *America*, but all of this class which are found wild, have sprung from swarms deserted from *Apiaries* in the settled provinces; there are none wild about the *Mississippi* yet, but a few years will certainly bring them thither; it is evident from this relapse of the bees in a wild state, and their prodigious thriving, that honey and wax may be made very considerable articles.

LI. *Ginger* will certainly be found profitable if introduced into the southern parts of *East-Florida*.

No doubt many of my readers begin to think, I have swelled this account beyond all bounds; I beg pardon for intruding on their patience, if I have done it, but the inquiring philosopher will, I suppose, find that considering the variety of climates and soils, I might have gone on, and reasonably enumerated a number of others; my earnest hopes are that these loose hints may prove beneficial to a continent, which has so great a share in my warmest wishes.

I have every where avoided describing expensive ways of culture, such are too much obtruded on gentlemen who begin the noble science of agriculture; men are discouraged by following such precepts; in those hints which are new I speak from a degree of experimental observation, in others I have chosen such methods as are in use and least expensive, I have shunned a mere *theoretical* narrative, because I utterly hate it: unpractised methods often appear plausible, even when they will produce effects diametrically opposite to what were intended.

It is true, that the pine barrens (so called) will in all appearance not be soon wanted, especially in *West-Florida;* but I think that for those, who are not possessed of hammock land, a hint for the improvement of these pine lands may not be amiss; we see a surprising vegetation effected on some of them: complaints of barrenness may sometimes be removed by meerely adapting the soil to its proper use; a man who has a good piece of river land, bordering immediately on these barrens, without any intermidiate oak or hammock soil in *Georgia, Carolina* and *East-Florida,* thinks himself almost always deprived of means to raise *Maize,* and will think his staple land of less value on that account: let such a person well examine his pine land, perhaps the clay or marl lays at a small depth below the surface, in that case, by mixing the soils, you may improve them so as to be fit for any purpose. None of these lands are so absolutely barren but they will produce sweet potatoes, or pumpkins; I have seen *yams* come to good perfection in *Georgia* and *East-Florida* in such lands; in *West-Florida* I saw them pretty forward in the navy garden of *Pensacola,* which is comparatively a meer beach. Peach orchards will do here as well as in the richest soils; the worst kind of this land will produce rye to advantage, even twice in one year; the *sesamen,* or oily grain, a profitable article if well attended to, will grow kindly in them, and they produce crop grass (a good pasture) in abundance, by merely turning them up often: for these reasons then a man ought never to be too precipitate in pronouncing a piece of ground *absolutely barren;* we have in these provinces no undrainable boggs, no mountains, and very few stones to deal with.

The manner in which cattle are now kept in the southern colonies is unprofitable; twelve or sixteen cattle might be with a little attention made to yield more profit in a dairy, than flocks of three or 400 do now, with all the labour and time at present bestowed on them; the practice of letting calves suck so long as they now do, is contrary to the practice of all ages and countries, except among the Spaniards; who keep large stocks for the sake of their hides and tallow only; and as in the southern colonies the people seem more and more to neglect keeping a few cows at home for the dairy, and hides and tallow cannot be made large articles in trade, by means of small stocks of five hundred or six hundred only, I can hardly see the intent of keeping such stocks of cattle, except for exportation alive, as the stocks of horses are kept; for in the article of beef, cattle can hardly yield profit where the *Carolinian* or *Georgian* method of killing at two, three, and four years old obtains, which is the cause of the badness of their beef.

Hogs are so profitable an article, and so easily made spontaneous, that it is a matter of the greatest surprise to me that no more are raised in Florida; especially as mast is very various, and in great abundance.

The bounteous hand of nature has here given us an animal, which, by experience, we know may easily be domesticated, whose fine wooll might yield good profit, and whose flesh is equal at least to our beef, and yields as much tallow; I mean the buffaloe: if instead of wantonly destroying this excellent beast, (*for the sake of perhaps his tongue only*) we were to endeavour its domiciliation, either by a pure breed, or by raising a spurious one with him and our common cattle, I think we could find our acount in it.

The *Moose*, or *American* Elk, found in the higher latitudes on the river, naturally leads a life so nearly approaching to a state of domestication, that I am often surprized he has never yet been attempted to be incorporated among the class of a useful tame animals on this continent. These and the like hints I think deserve attention, and I hope they may be of use. In a country where horses and cattle are in such abundance, and consequently cheap, and where human labour is so dear, more improvements in machinery ought to be introduced: it has amazed me often to see people so blindly mad as to ruin themselves, by obstinately persisting in building saw-mills in improper places, which has been here too often the case. The *Mississippi* indeed affords the means for mills at a small expense, but they are only of temporary use. In such a timber country as the *Floridas*, introduce the *Dutch* windmills, it is easily done; I have frequently heard, that they are unfit for this climate, but in my strictest inquiry not one reason appears against them, but that people by infatuation will run on a sand, where so many before them have suffered ruin by making dams in improper places.

As much as possible diffuse the hand hoe, particularly at turning up, and otherwise preparing the ground for seed; introduce the plough; the newest lands may be ploughed, and the hand hoe is only necessary to assist a little round the edges of stumps, &c. introduce likewise the horse hoe with the *Dutch* and *Suffolk* foot ploughs, to do the laborious work of the hoe in hilling corn up: in well improved grounds for rice use the drill plough, especially in countries where an ox of four years old may be had for forty shillings, and a serviceable horse for four pounds; all these would save great sums, and render it less necessary for the planter to depend on the labour of negroes.

In lieu of the aukward tool made use of to cut grain, I mean the sicle, I would advise the introduction of the short scythe and hook, called in *New-York* government *segt* and *mat hook;* and in case of thin crops the scythe and cradle: by means of these, expedition and neatness will become common, whereas now the crop stands sometimes on the field till it is damaged, unless a great number of slaves can be employed.

Threshing is performed in the most aukward manner imaginable, in the Southern colonies; to many the common flail is entirely unknown, and long poles, or a crooked cudgel made use of, and where the flail is used it is ill managed; for the large crops of rice, &c, I would recommend a machine which I remember to have seen used in *Holland,* as long as my memory can serve, and yet strange to tell, even in *England* the people were breaking their heads for the invention of a threshing machine no longer than four years ago; when the same machine had been known, and commonly used in *New-York* and *Pennsylvania,* near fifty years: it consists of a wooden cone, whose base is about two feet, or two and a half foot diameter, and whose length is from nine to twelve feet, it should be made of very hard wood, its surface is regularly fluted, toothed, ribbed or indented; the ribs are about fifteen in number, with as many grooves alternately between them, of about three inches thickness and depth; the small end of this kind of roller is furnished with a strong ragged eye-bolt and ring, and iron hoops to prevent its cracking, or the bolts drawing, this ring should be about six inches diameter in the clear, and be fixed on a post in the middle of the floor, which for twelve or fourteen inches up ought to be cased with iron, to prevent the ring from cutting it off by the continual friction: the sheaves being laid circularly round this post, the roller is by one or two horses or oxen drawn round in a constant circular process, one man attending to turn up the straw continually; this is more expeditious and cleanly than the trampling of it out with horses and cattle, because the thing is so contrived as to let the cattle walk without all, and it makes incomparably more dispatch than the flail can even with many hands.

In the common way of shelling *maize* there is also great loss of time. A man can hardly shell six bushels per day, whereas with the flail he will shell twenty in an hour, and the fanning mill will soon clean it; many more improvements will in time be wanting, but the above are now absolutely necessary.

Thus far I have given a detail of the agriculture practised, or practicable in those climates; during the course of my collecting materials for this narrative, I have sometimes been beyond all measure vexed, and at others I have been obliged to laugh at the silly notions, whereby *England* is deceived in her ideas of *America,* occasioned by some foolish writers, who have raised some absurd hypothesis in their own brain, from whence they deduce as crooked *theories* as ever entered the thoughts of mankind; thus writing without experience, they lead a parcel of blind copiers, as it were, in a string, as if intended on purpose to deceive mankind; and what is more strange, some people so implicitly believe these ingenious writers, that I have seen them after coming here obstinately follow the precepts of these guides, though in direct opposition to what they see to be the practice of their neighbours, who have been taught by experience. How sillily has one vented an opinion, that by clearing the woods of *North-America,* we would ruin our temperate colonies, and turn them into cold uninhabitable desarts, thereby bestowing the climates of *New-England,* on *Carolina;*† with more such fancied evils, notwithstanding that even when he wrote, *Canada* (by the *French* only deemed fit for a fur-factory) began by *English* industry to become a grain country; so that he might have known better, and now that same country even exports wheat; pray what is the reason of this but that the clearing the land causes the snow to remove sooner

†Present State of *Great-Britain* and *America.* 8vo. 1767.

than it formerly did? Under the head of sugar in this work see more of this.

Just such nonsense he vends, when he attributes the want of wood in the plains of the north western parts of *America* to barrenness, when experience has taught every one that they are fertile; how well does it confirm this genius's opinion when we see without deception, that the nearer we approach those plains to the west, the more temperate is the climate! At *Ilionois*, in latitude 40, snow seldom lays three days, and cattle are out all winter; in the *Nadouessin's* country, latitude 45 and 46, above two thousand miles from the Atlantic, snow rarely exceeds three inches in depth, and the wild rice is found spontaneous in amazing tracts. Nor is his argument of the cold being more felt in a field than in a wood of any force; let him remember that snow lies a month, or even six weeks longer in a wood than in a field. This mans dares to tell us "we talk from experience." So it seems, when he says the badness of pasturage in the southern colonies renders it impracticable to maintain stocks of cattle: was it my business I might enumerate more such shining authors;† but as it could serve no good purpose I omit it, and only beg leave to hint that such blind guides ought not to be too implicitly followed.

In the above account I have inserted some articles which might possibly have come more properly under the head of manufacturing, such as *indigo, pitch, tar, turpentine*, &c. and even *rice*, but their near connection with tillage has made me insert them there, rather than in the following part of this work, where I have dedicated some pages to hints concerning manufactures.

Reducing the rough timber into boards is now the most common branch of business in the southern colonies, that comes under the name of manufactures: a pair of negro sawyers are taxed to cut an hundred feet per diem, this the industrious ones often perform by two o'clock, or sooner, they are then suffered to cut lumber for their own use the rest of the day, or do any other business, for which their masters pay them; these hundred feet *per diem*, are six hundred per week, or thirty-one thousand two hundred *per annum*, which, allowing them to be only inch boards, and to sell at six shillings sterling

---
†There is one so very cunning as to have discovered the labour of our slaves here to be dearer than that of hired people. Amazing wisdom! See the Appendix to the Interest of *Great-Britain*. 8vo. 1760.

per hundred, is £. 93, 12 s. *per annum*, a noble benefit! which in thicker board, scantling, and ranging timber is much greater; the owner of such a pair of slaves lets them out at £. 60 *per annum;* from hence we may gather how necessary the introduction of saw mills is to lower the price of this article.

Hogshead staves of white oak are made by what are called gangs of people; a stave making gang consists of five persons, a feller of the timber, who cuts down the trees, two to cross-cut them in proper lengths, a river or splitter, who rives them with the fro, and the fifth is employed in shaving them; this gang makes five hundred staves each day, which are supposed to be worth 30 s. is 6 s. *per diem* for each man.

Shingles of cypress and white cedar, which are made nearly in the manner of staves, are sold at about 10 s. per thousand, and one gang may very well make three or four thousand per day.

But the grand manufacture to be made of timber here, is SHIPPING, for this purpose no country affords more or better wood; live oak, cedar, cypress, yellow pine, are adapted by nature to this. O! how just is every *Englishman's* reason for cursing the late peacemakers, when he reflects upon the fatal mistake of leaving the isle of *New-Orleans* in the hands of the *French* and consequently of the *Spaniards!* and when he sees them building such fine frigates as they did last year on that island; to add to the misfortune they leave their own timber and cut it off of the *English* land, about the lakes, for present use; might not *England* herself infinitely rather build ships of war, and sell them to her enemies, and so make profits of them, than to be obliged to behold this with supineness? When we recollect the amazing distance into the heart of the country to which the *Mississippi* and its branches gives us access, and where we may build ships of two hundred tons at least; when we see the iron mines dispersed up and down the country; when we recollect the possibility of producing the most immense quantities of hemp and flax; when we behold the more than rapid increase of mankind there; and finally when we survey the harbours, *Charlotte, Tampe, St. Joseph, Pensacola,* all proper for the admission of ships of rank, besides others we know not yet; what a field is open here! what a prospect of power and grandeur seems to be already welcoming us! no country had ever such inexhaustible resources; no empire had ever half so many advantages combining in its behalf: methinks I see already the *American* fleets inhabiting the ocean, like cities in vicinity!

The manufactory of iron will undoubtedly be very great here; every part of the western province at least abounds with it; no doubt other minerals are hidden in the bowels of the earth, but except lead and antimony they are not yet found.

The article of *Potashes* of various kinds have been mentioned under the head of *Mangroves* and *Salsola;* I shall only repeat here that no country can make these commodities equal to *East Florida;* that province can undersell and supply all others of the earth, when we consider that *England* takes £. 50,000 sterling worth out of *North America* alone, where people can not make it either so good or so cheap, it is spur enough to encourage some person to begin this manufactory in that country, but be aware of doing it in a manner clear of *Steven's* and other expensive schemes.

Deerskins though none are manufactured here, yet, being a great staple commodity already, they deserve a place where, was it but to remind my reader that the *Floridas* are watered by all the rivers by which peltry is like to come in future, and that by a little attention and pains taken with that trade, these two provinces, particularly the western one, may lead it down their own channel to that degree, that every individual skin (got to the westward of the heads of the *Ouabache*, and so up to the western branches of the *Mississippi*, and to the southward of *Ohio*) must, as they certainly ere long will, center in the gulph of Mexico.

When we consider the vast increase of stocks here, and the ease of *maintaining* a stock of a thousand or fifteen hundred cattle, and four or five hundred horses, and the very great difficulty attending the northern colonists in keeping up a stock even of no more than twenty or thirty head of black cattle, it must be evident, that tanning of leather will be a great business here; even allowing the number of *American* inhabitants to be no more than three million (which number the late systematic writers seem determined to confine us to) and allowing two and an half millions of these to wear about six pair of shoes per annum, and that in *America* about ten millions of these are annually made of *American* leather, which is more than likely to be the case; the town of *Lynn* alone exporting about three hundred thousand per annum, there will then remain about five millions to be supplied from *England*, even this leaves a large field open, and if we are industrious not only that gap will be stopped, but we may export tanned leather, if *England* does not want it, or will not take it, other countries will; consider ye *Floridans* what an extensive manufacture of *Chamois leather* your lazy neighbours of *Campeachy* carry on already.   :

When we consider that the northern inhabitants clothe at least seven eights of themselves with cloth, that costs them hardly any thing, but their industry, when we ponder well the indefatigable pains taken by the *Acadians*, settled on the *Mississippi*, to array themselves in every respect with the produce of their own fields, and the work of their own hands, when we see the immense quantity and variety of raw materials afforded by this country, such as wool, cotton, silk, &c.. heretofore enumerated, must we not be inspired by the genius of frugality, to make use of these blessings of the bounteous hand of nature, and to imitate these industrious people, whom Providence already suffered to be driven out of the land of their fore-fathers by the iron rod of arbitrary *Mars*, and caused to settle in our neighbourhood? If we learn by times to know, that we enjoy all these blessings, we shall have but little use for importing any thing but money.

Pork and beef will shortly be worth attending to on the *Mississippi*, as is evident from the quantities of salted wild beef, already exported by the *Spaniards* to the *Havannah*.

An article little known, perhaps not thought of, next claims our attention, I mean the extensive fisheries in the power of these colonies, with which they now supinely, not to say stupidly, allow the Spaniards to run away. The whole of the west coast of *East Florida*, is covered with fishermens huts and flakes; these are built by the *Spanish* fishermen from the *Havannah*, who come annually to make one or two fishing voyages on this coast, to the number of about thirty sail, and one or two visit *Rio d'Ais*, and other places on the east coast of the penninsula, they arrive about the latter end of *August*, and continue coming and going till the end of *March*, their first care is to prepare their nets, and to build a hut, or refit the old one; then they new furnish their flakes or stages with new strings of silk grass to the wooden hooks by which the fish is to be hung up to dry; their nets and other apparatus of lines, &c. are all made of silk grass likewise; and as they fish upon shares, each man furnishes his piece of net, number of lines, share of salt, and *quota* of provisions for the voyage, or if not able to purchase all these (which is seldom the case) the proportion of such person is furnished by the owner or captain, who deducts their value from his share, when he receives the dividend of the neat proceeds of the cargo; the distribution of this dividend is generally as follows: the vessel draws one third, the *patroon* or master, two shares of the remaining two thirds is divided into as many shares as there are adventurers; reckoning two boys for a man: their charges are little more than the wear and tear of the vessel, the purchase of salt from the King, and the duties of entry for the fish, which last I think is two and an half per cent, with these people a profitable trade might be established; they have often told me, that in case they could find *Englishmen* on the coast, who would furnish them with salt, they would not purchase any more than what was absolutely necessary to cover their license, for these very men are obliged to go to *Key Sal*, to gather the salt there, and bring it to the *Havanna*, where they are obliged to deliver it into the King's ware houses at three *rials* for a *fanego*, a measure containing scarce two bushels, and when they prepare for the fishing voyages, they are not suffered to take any salt elsewhere but out of these warehouses, where they are now constrained, by their most gracious Sovereign, to pay one dollar and an half, or twelve *rials* for the identical salt he took from them for three; provisions and dry goods might also be sold to them to advantage, but it would be necessary, that they were made certain first of finding such vessels on the coast, for without such a certainty, they would not venture to come without salt, nor would they bring any money; the same vessel that carried salt and dry goods might here drive an advantageous trade with the parties of lower *Creeks,* who come to hunt in the same season, in this part of the peninsula, and always get great quantities of skin, she might likewise get a load of fish, which would answer well in the *West-Indies*, as the fish from hence always fetch a considerably greater price in the islands, than the fish from *Newfoundland*: The principal fish here, and of which the *Spaniards* make up the bulk of their cargoes, is the *red drum,* called in *East Florida* a *bass*, and in *West Florida carp;* the *French* call them *poisson rouge;* this is in those provinces a fine fish, although in the northern colonies they are generally poor. They also salt a quantity of fish which they call *Pampanos,* for which they get a price three times as high as for other fish. A few *soles*, some *sea trout*, and the roes of *mullets* and *black drum* make up the remainder of the cargo; some oil from the liver of *nurses* and *sharks* is also carried; of the sound of the *sea trout* a glew is made by only drying them, which is a perfect and fine *ichthyocolla*. The roes of *mullets* and *black drums* are put into a pickle for about a quarter of an hour, then taken out and partially dried in the sun, then pressed between two boards; afterwards exposed upon a hurdle in a small hut to the smoak of the inner part of the ears of corn, which is properly the receptacle of the seed, and called the *cobs*. These roes the *Spaniards* are very fond of, and use them instead of *cavear*. From this account my reader may judge of the worth of this fishery. These fishermen make likewise no inconsiderable quantity of *shark oil*, and it is well known what number of *turtle* the *Providence* people catch on the east coast.

The *Myrica*, or candleberry myrtle grows in great abundance on the sea-shore of both these provinces; the wax produced by it is too well known to need description; it is manufactured by throwing the berries into a hot bath, to discharge them of their wax, which is skinned from off the top after the water grows cold.

Travelling through the uninhabited part of the woods, falling to the share of every person, who arrives at present in *Florida*, it cannot be amiss to say a few words about it: whatever you go by water or by water or by land it is most necessary to carry with you powder, shot, and a fowling piece; provide yourself with rice, or such bread as you like best to use; a hunter is necessary and utensils for dressing provisions; always before night allow yourself at least one hour's time to prepare fire wood, for this must be used in summer as well as winter to rarify the air round the camp; always lay with your feet towards the fire, and you are out of danger of catching cold; if you are in a country where warring savages resort, keep up a large fire all night, and be sure to put some hats on poles near the fire, this will protect you from their attempts, unless they are at war with ourselves. A bear skin on the ground to protect you from its dampness is very necessary; always choose an elevated spot, and if you have horses chuse a situation near a cane branch, or where the woods have been lately burnt, this will effectually prevent your beast from straying. As the diet is generally roasted meat remember my former caution against *spits* got out of a low ground; small reeds are excellently adapted to this purpose: a small hut covered with thatch of *palmittos*, or bark of trees, is always preferable to the lumber of a tent; such an one, sufficiently large for two men at least, is easily set up in a quarter of an hour; if you are in a river be sure of fastening your boat in a still water or eddy, and be cautious not to fasten it with a wythe, which is too common a practice with many; this precaution will prevent your boats going adrift by means of a beaver's appetite. If you are where the grass is long, burn a space of ground before you make your fire, this will save your baggage, which many have lost by neglecting this; if in a part of the country where musketoes are plenty, have a close covering, called in this country a *Bère*, and made in form of a musketo net, to put up over your bed, suspending it by two stakes; and lastly encamp not near any old trees, but as much as possible shelter your camp against the wind; by carefully observing these hints, you will not find travelling through the *American* desarts so uncomfortable a business as it may generally appear to be; as for the dreadful stories told of wild beasts, believe me they are vain, no animal is yet found in the *North American* wilderness ferocious enough to come within sight of a man, if the wind wafts the air of a human body towards the brute; if it does not, he will not venture nearer than in sight; but should you wound a bear, or the *American* panther, so as to disable him from flight, he will prove dangerous, but remember, if you tread even on a worm, he will turn up his tail at you, as it were in his own defense.

I shall now say something about the most adviseable method to be taken by emigrants, who intend to transplant themselves into this part of *America*, from the more northern regions; first I would observe that if destined for *East Florida*, it is absolutely necessary to carry at least one year's provisions in flour, maize, pork and beef; if to the eastern part of *West Florida*, the same step is to be taken; but if to the *Mississippi*, provisions barely for the voyage will be sufficient, they being always to be had there in plenty at or nearly at the prices beforementioned! if the person be able I would advise him to purchase negroes in the northern provinces, and to carry a few more than he intends for his own use, the profits on the sale of four or five will nearly defray his expences; carry no white servants, unless you have a mind to colonize a large tract of land, and this has never yet turned to account; I will suppose two instances of what in my opinion is the most elegible method of settling people of different situations in life.

Let one man be possessed of two thousand five hundred Dollars in money, and we will suppose him living in *Rhode Island*, or in any other part of *New-England*, or *New-York*, *New-Jersey*, or *Pennsylvania*; and allow him to have a wife and four children, with two house slaves, in either of these colonies; he may purchase eight good working slaves for twelve hundred dollars, out of his two thousand five hundred; about four hundred dollars more will buy four young girls or boys for which he will, in *Florida*, find ready sale, with 80 per cent advance, but they ought not to be under twelve or thirteen years old; his next care is now to get a vessel, we will suppose her of sixty tons burthen, such a one as may generally be had manned and victualed, at two dollars and a half a ton, per month, from any of these provinces, the planters provisions and implements for himself and family will cost about three hundred dollars, allowing him well provided with every necessary for eight weeks; thus we may suppose him arrived in the *Mississippi*, with six hundred dollars left him in money: here we will leave him a while, to pay a visit to a poorer family, we will allow the head of this to have as many cattle, horses, hogs and superfluous implements as may raise him four hundred dollars, for a poorer man should not attempt to go as a planter, he may as an artificer, or tradesman, we will also allow him to have an equal share of matrimonial blessings with his richer neighbour this man must lay out about sixty dollars in provisions, and about twenty-five for plantation tools, his passage money will amount to fifty dollars more, we may suppose him possessed of one good able negro, besides his money which may be worth an hundred and sixty dollars; thus he may arrive at the *Mississippi* and have ninety dollars clear money after obtaining his land and stock; here by the way let all who intend going to this country endeavor to arrive about the latter end of *November*, in order to clear some land and build some place for shelter. White servants will never turn to account, there being so many idlers already imported on wrong plans, that you can carry none with you who would not in three months time think it very hard to be obliged to call you master, in a country where the most dirty vagabond you can hire at six Dollars per month, would think his honour touched by hearing any body call him to you with less civility than, *Sir, your employer would be glad to speak to you.*

    I will now throw this matter into a clearer light, by placing it in tables, in which the reader may see it at one view.

	£	s	d	Dol.
The first men being possessed of				2500
We will suppose him to prepare himself for the journey in New-York, where he purchases 8 working slaves, at £. 60 per head........................£.	480			
4 young do. for sale, at £. 40............................	160			
Allowing his 14 slaves, including his 2 house-slaves 2 lb. of rice, and half a pound of pork per diem, for 60 days passage; this will amount to about three tierces, containing 1700 pound of rice, which at 24 s. per Ct. amounts to..................................	20	8		
2 barrels containing 420 lb. of pork, £. 4 10 per barrel	9			
For his white family's use we will say, 2 tierces of 360 lb. of bread, at 17 s. per Ct.............	3	1	2	
1 barrel of 180 lb. of flour, at 22s. per Ct................	1	19	6	
1 barrel of choice beef, £. 3................................	3			
½ barrel of choice pork, 50 s................................	2	10		
50 lb. of hams, 1 s. per lb....................................	2	10		
a firkin, 60 lb. of butter, 1 s. per pound....................	3			
2 large shoats, 10 s. a piece................................	1			
2 sheep, 15 s. do	1	10		
1 dozen of geese and turkies................................	1	4		
4 dozen of small poultry....................................	2	8		
coffee, tea, and spices..	3			
3 loaves, 30 lb. of sugar.....................................	1	10		
25 lb. of muscovado ditto....................................		16		
15 bushels of corn, to feed his stock, at 3s. per bushel	2	5		
small expences, for greens, &c...........................		18	4	
Total in N. Y. cur. amounts to..........................	700	0	0	1750
			remains,	750
we will suppose him to carry one half puncheon of Jamaica spirits, and one half ditto of common rum, Dollars	50			
and to lay out for half a pipe of good Madeira, to serve him for his first year's stock.............	100			
				150
There now remains as above in his possession, at his arrival: of this he must			Dollars	600
pay for the charter of the vessel............................				300
			Remains Ds.	300

Suppose his 4 young negroes sell at an average for 150 Dollars each (which is not a high price) this amounts to..			600
		Dollars	900

In the above account no mention is made of furniture, because a man possessed of such a sum of money may naturally be supposed to be superfluously provided with that article, and by selling a part of this, he may supply himself with others, that have a peculiar reference to these climates, and are indeed not to be dispensed with: axes, saws, spades, hoes and other implements of husbandry perhaps necessary to the amount of sixty Dollars, may be likewise supposed to be purchased out of the sale of his superfluous furniture.

Since the writing of this part of the manuscript, the new mode of parcelling land out in small lots, and selling them at auction, was introduced into the colonies. This scheme, it is said, was first projected by Sir *James Wright,* Baronet, Governor of *Georgia.*

On what terms lands are to be had now since the shutting up the land-office is yet impossible to tell; but the undermentioned was the state of this business before that event happened, with the fees as they stood regulated in *South Carolina, Georgia,* and *East-Florida.* In the western office I was never conversant enough to know the real state of this matter; but I have seen some of Mr. *Durnford's* bills for surveys calculated at 50 per cent. higher; for what reason I know not, for surveying in *West Florida* is attended with less difficulty, than in either of the others.

Petition, warrant, and precept................................		11	7
To the surveyor general, and his deputy, each 2s3 per 100 acres; the number of acres for the above named family would amount *honestly* to 850 acres, which makes at 4s6 per hundred....................................	1	18	3
Suppose that 1000 acres were granted besides on purchase, this would cost 10s. per hundred........................	5		
The above-mentioned expences of petition, warrant and precept on the last 1000 acres, is...........................		11	7
Survey of 1000 acres, at 4s6 per 100........................	2	5	
Suppose the land 150 miles distant from the surveyor's residence, he being allowed 6s9 for every 20 miles out and in, except for the first 20 miles, at 6s9 per 20 miles out..	2	3	10½
The same 130 miles at 6s9 per 20 miles home............	2	3	10½

Allow the surveyor 14 days to go and come back, he has four men with him to row in his boat, or carry his provisions and instruments, if by land these same two men serve as chain-bearers, and two as blazers, they are generally paid for at 2s6 *per diem*, each........ 7

Provisions for 5 men for 14 days are to be had plentifully, and good, at one shilling *per diem*, rum included 3   10

Return of two precepts, plats and certificates by the deputy surveyor, 6s9 each... 13   6

Return of the two warrants, platts and certificates by the surveyor general, to the attorney-general, 6s9 each 13   6

Two *fiats* by the attorney-general, 6s9 each................ 13   6

Governors and Secretary's fees, about   8

                Total Sterling *L*.................... 35   4   8

      Which £. 35 4 8 Str. (in consideration whereof the planter is now in possession of 1850 acres of land,) are at the exchange of 4s6 per Dollar............ Dol. 155

Building a comfortable neat timber house, out-houses and negro hutts on this land, will cost about................. 250

Maintenance of the planter and his family for 4 months, during which time all the above business may be done and the house at least rendered habitable, the expence of travelling to Pensacola and back to Mississippi included.. 150

                                       Deduct these   555
                                             from   900

and we will leave him now in possession of his land, house, two negroes, and cash to the amount of............ 345

      Suppose the month of April now set in, and his negroes in these 4 months to have cleared him 20 acres of land, of which he plants 10 with *Maize*, pease, &c. 3 with rice, and 7 with indigo, from this he may reasonably expect at the most moderate calculation on the Mississippi lands, about 400 bushels of *Maize*, and 300 of pease, with perhaps about 500 pumpkins, about 150 bushels of rough rice* and 400 lb- of indigo by the beginning of November; we will suppose that in this same interval he has purchased

Six milch cows at 12 Dollars each... 72
Six hogs   4   24
A stock of poultry   10
A horse   30

---

*It is necessary to observe here what I forgot before, this is that about three pecks of rice are sown on an acre.

Having lived at home between April and November, we may reasonably suppose these 7 months to have been less expence to him, and to have cost for maintenance no more than the first mentioned 4 months or...	150
Expences of building his indigo vats...	40
A boat for plantation use..	15
	341 Dol.

We must suppose him to have converted one of his house negroes into a field slave, which will make his working hands 9 in all; by no means too small a number to clear and cultivate 20 acres for the first year. We have now indeed seen him expend nearly all his money, but the year is come round and he has now 400 bushels of corn, 200 of which together with the pumpkins he must reserve to feed his negroes and stock with during the next year; thus remains 200 bushels of *maize* for sale at 2 ryalls per bushell as before said is.. 50 Dol.

Of 300 bushels of pease (150 reserved for his use) remain 150 for sale at 2 ryalls also... 32½

150 bushels of rough will make 3500 lb. of clean'd rice, which his negroes manufacture during the first winter by hand; this is about 17 barrels of the Mississippi measure, of which 2 being kept for his own use, there will remain 15 barrels for sale which at 2½ Dol. per barrel is.. 37½

400 lb. of indigo at the price it bore in November and December 1772, is 400

Thus we may suppose for this first year his live stock in a thriving condition, his stores full of provisions, and his cash amount to .... 520 Dol.

He may at present reasonably be thought to increase his working negroes to 10, and to add 10 more acres to his cleared ground, besides improving last years clearing. I will now leave this man and give some advice to his poorer neighbour — I appeal to very man who will be candid and is really acquainted with the country I speak of, whether my calculations are just or not, and whether my balance of the first years work is not exceedingly moderate: in regard to a buck or jockey, who expecting to find this new country on his arrival full of Vaux-halls, Ranelaghs, and New Markets; who just comes to scamper over it, and when he finds his expectations failing takes a look at the woods, and at the moderate life of the new planter, then turns up his nose, takes a pet and goes off, puffing of a "d--n the the country," he can at best pretend that he was there in a dream; he is no judge, he has no business with serious books on the present subject, at least I write not for him.

	£	s	d		
The poorer planter we have before allowed to possess				400 D.	
The provisions necessary for his voyage for his negro and all included, to be bought at New York as before,					
2 tierces of bread £.	3	1	2		
1 barrel of beef	2	5	0		
1 barrel of pork	4	10	0		
25 lb. of ham	1	5	0		
1 firkin of butter	3	0	0		
4 small shoats	2	0	0		
2 dozen small poultry	2	8	0		
25 lb. muscovado sugar	0	15	0		
10 bushels of maize	1	10	0		
coffee, tea, and other small expences, rum included	3	6	0		
	24	0	2	equal to	60 D.
				remains	340
his passage money					50
				remains	290 D.
I will suppose the plantation tools he wants may amount to					25
				remains	265 D.
The expences attending a survey of 400 acres of land (150 miles from the surveyors house,) the quantity this family was formerly intitled to, are as follows, viz.					
Petition warrant and precept as before....£	11	7	Str.		
Surveying 400 acres at 4s6		18			
Milage	2	3	10		
Allow the surveyor 12 days with four men, but at the same time the planter may be supposed to have a son, being an able lad, who together with the negro make out three, thus remains only one to pay 2s6 per diem	1	10	0		
Provisions for 5 men 12 days rum included, 1s *per diem*	3	0	0		
Return of precept warrant, plat certificate and fiat	1	3	0		

Governors and secretarys fees...........	3	0	0
L	12	3	8 Str.

Equal to 52 Dollars.

Himself, his son, and negro employed between November and April, in building a comfortable house of square cypress timber, dove tailed, consisting of two rooms and a loft, together with a corn-house, and in clearing 8 acres of land, during which interval the maintenance of his family may be supposed to cost........................... 80 D.
Allow that he buys in the same interval
a cow .... 12 D.
2 sows, 4 D... 8 D.
a stock of poultry... 6 D.

158 D.

A horse ....
A canoe we may suppose built by himself, iron work excepted... 2 D.

175 D.

Thus we see him in possession of 400 acres of land, a small plantation stock, and........ 90 D.
Suppose his clear land planted in the following proportion, viz. (in cash.
5 acres with maize, pease and pumpkins,
1 acre of potatoes,
2 acres of indigo,
It may reasonable be supposed to yield
      200 bushels of corn,
      150 do. of pease,
      300 pumpkins,
      100 lb. of indigo.
During summer, his son and negro attending the land, he may be usefully employed in squaring timber, or in some handicraft he understands, this together with the increase of his little stock, and about 60 dollars well laid out, we may reasonably think will maintain his family from April till November .... 60 D.

Thus the remainder of his c a s h now amounts to .... 30 D.

We must allow himself to have made his small indigo vats,

He has now for sale 100 lb. of indigo........	100 D.
120 bushels of corn 2s..............	30 D.
100 do pease, 2s.........................	25 D.
	155 D.

Thus he may very justly at a moderate computation be said to remain at the first years end in possession of 400 acres of land, a comfortable house, improving stock, and fruitful farm, besides in cash....

185 D.

Let it always be remembered, that the same calculation I have made for the Mississippi, will nearly answer for the *Tombechbe* or Mobile river too; but if you intend for the eastern province, or for *Pensacola*, be cautious and bring provisions along with you to last you at least the first year. It may not be amiss to recapitulate in a general view the present exports of these provinces, together with the possible additions, which to all appearance may be made to them as mentioned in the foregoing pages.

1st. Articles already exported from these provinces.

Indigo,
Maize,
Rice,
Tobacco,
Mules from Spanish Louisiana,
Pitch,
Tar,
Squared timber,
Cedar posts and plank,
Cypress & pine boards,
Plank of various woods,
Scantling,
Staves and heading,
Shingles,
Sassafras,

Indigo seed,
Salted wild beef,
Carravances,
Live cattle from West Florida,
Hoops,
*Dried salt fish,
Canes,
Oranges,
Deer-skins,
Peltry,
Myrtle-wax,
Pacan-nuts,
Raw hides,
Buffaloe tallow,
Bears oil.

2d. Articles, some of which are already naturally found, and others, which to all appearance will soon be introduced, and must become staple commodities in these provinces,

Wheat and wheat flour,
Rye,
Hemp,
Flax,

Bees-wax and honey,
Silk,
Oil of olives,
Cotton,

*This article and Cavear with some Ichthyocolla are already exported by the Spanish fishermen, the two first in very considerable quantities.

Tanned leather,	Oil of Benni,
Salt beef,	——— of ground nuts,
Salt pork,	Cavear,
Ichthyocolla,	Turpentine,
Wine,	Oil of do.
Madder,	Rosin,
Rhubarb,	Gum elemi,
Jalap,	Storax,
Pot-ash,	Figgs,
Barilla,	Raisins,
Safflower,	Drugs of many kinds.

3d. Articles which we have the greatest reason to believe will in process of time become staple exports from these provinces; these are,

Sugar,	Coffee,
Rum,	Pimento,
Ginger,	Cacao.

4th. Articles of which it were to be wished that the introduction would be attempted soon, under some encouragement,

Tea,	Cochineal,
Arnotto,	Cork Tree,
Orchil,	Spurious breed of the buffalo.

I have used my utmost endeavours to collect materials for ascertaining the enumerated exports of Carolina and Georgia for the last two years, but to no purpose, except a paper of the best authority wherein the state of the exports of the last province are recapitulated, and their value ascertained; this paper was printed by James Johnstone, in Savannah, and is republished in this work, with a view to convince my reader by the most undoubted matter of fact, that the province of Georgia has advanced the value of her exports in about seventeen years time from £ 15000 to £ 121000 sterling per annum, before the 1st of January 1773, as shewn in page 104 of this volume, this together with a consideration that in the year 1768, when Georgia was still poor, that province and Carolina were thought to increase the wealth of the British nation, near a million sterling* will make us reasonably judge that these two provinces under consideration, (one of which is at least equal to Georgia, and the western one infinitely superior to both Carolina and Georgia in point of the quantity of fertile acres) may in 10 years time do as much, provided that the course of nature be not forced into another channel, when either by fraud or violence means will be found to deprive the inhabitants of the Floridas of the benefits which have accrued to Carolina

---

*Political essays 4to London 1772. p. 359.

and Georgia, from that law of nature and of God, the *lex agraria*. Commerce is so effectually necessary to all people on earth, that none can be said to subsist without it; it is in this part of the world chiefly necessary on one account; this is to acquire the means of power to defend the country by riches, which in all probability we must expect to do chiefly ourselves, if we may judge from the policy of destroying the forts *Toulouse* and *Tombechbé* among the chain of forts that defended the colonies from the inroads of the savages; it may also be of use to maintain the poor, for of these there are already no inconsiderable number who not being brought up or used to tillage are obliged to become fishermen and hunters, contributing to support the naval power of Britain may likewise be called a third important motive to urge us on to a promotion of trade, but this last cause of incitement is yet in embryo. Let us endeavour to build vessels ourselves, let not our trade in this country (where ships may be so easily procured) be confined to selling our products at home to people from abroad, who will come to fetch them at pleasure; if we do, what must become of the hopes of ever extending our present limited settlements! consider that on this point depends the increase of mankind, and the welfare of every country.

Never let us suffer monopolies from abroad; observe with what a fatality they are attended, see what a large cultivable tract of America lies waste through the influence of a mean dirty company, who do not annually trade to above one fourth the value of what some private merchants do both in England and Holland, and who have ever with might and main in private opposed the pursuit of a business* the very effecting and encouraging of which was the fundamental intention and condition of the charter which intitles them to be at present such a set of illegal regraters.

When we consider again the effects of the establishment of European factors on the trade of Virginia, we shall likewise see the effects of monopolies to be pernicious; for this very scheme, however strange my assertion may appear, is a monopoly in disguise; keep such *mangonizers* from among us; it is true, some few merchants will be at first possessed of the chief of all the trade, and thereby (if they are not more virtuous than the generality of mankind at present are) have it in their power to do much mischief to commerce, but let it be remembered, that as long as they are not factors from Europe there

---
*Discovery of a north west passage.

is room for others to come in, whereas if once European factors take place, who have the sale of the European manufactures consigned to them, with orders to keep the price of our valuable products within their own stated limits, it is also done with us; commerce will be so cramped that the innumerable quantities of profitable and fertile acres of this wide extended waste of new world might as well be converted into a sea, and think not these cautions premature; there is no want of designing men to set such a scheme on foot to the prejudice of these new countries, and too many ungrateful sons of America itself, would be found to turn their acquaintance with its commercial interests to advantage in gratifying a desire to have a bite at her vitals. I would not be thought, notwithstanding this, to contradict the advice I gave in page 74 of this volume; that scandalous licentiousness which so greatly prevails among the present traders, is so great an evil, that it is become highly necessary a less one should be introduced to effect the destruction of the greater, by monopolies of the trade of each particular savage nation being granted to different men, who would go to reside in each of these different nations; nor should they have this stretch of favour bestowed on them by the community, unless they are laid under proper legal restraints, and until such time, as an amendment in the present distracted state of this great and profitable branch of trade might be effected.

It would scarcely have been necessary to say so much on this subject, were it not that new countries are so open to an introduction of novel schemes, which will be found of pernicious consequences when it is too late; I say it therefore once more, oppose all schemes that may have a tendency to introduce a monopoly of your products, for a conclusion, hear from me the opinion of the greatest statesman Europe ever produced;* he says that it was the opinion of the greatest statesmen (undoubtedly after him) that had the rulers of that respectable union in 1609 dealt in the same manner with every branch of trade as they did with that of the East and West-Indies, *not one tenth part of the inhabitants of that opulent country would have been able to live and earn their bread;* and Holland would have been ruined. Another writer poetically exclaims, "this (meaning monopolies) is the spring from whence misery overwhelms the people."† In short, to me it seems nothing can be more evident than

---

*Jan De Witt, counsellor pensionary of Holland.
†*Hac fonte derivata cladis in patriam populumque fluxit.* Thoughts of the present state of trade in India.

that the establishing a rational commerce, in an equitable manner, should be a principal aim at the same time with the establishment of agriculture, and introduction of profitable staples in a country so peculiarly full of advantages for the promoting of these grand objects, as the western province especially is.‡ The particular branches of

‡I am here tempted, although not the least appearance of such a monopoly exists yet, to lay before my readers the manner of government at Surinam, a colony of people who are beyond contradiction possessed of *real liberty* in all its extent, and whose political history (notwithstanding their vicinity to Great-Britain) remains absolutely a dark labyrinth to almost every individual of all ranks and classes, high or low, in the British dominions, notwithstanding the knowledge of a pamphlet, called *Observations upon the Netherlands*, whose author§ (I ask pardon of his respectable memory) has only made some crude and short remarks, in which he has in the most evident manner, published his thorough ignorance of the state of the government he pretended to explain. This mode of administration would never have been erected by my countrymen over their plantations, was it not that a *monopoly* was at the head of it, what I offer here is a faithful translation of some papers authenticated in the utmost extent of the word.

"The *governor*, who at the same time is *colonel* of the military, (not the militia only) has the supreme command over the colony, as well in civil as military affairs: he is appointed by the society; but his appointment must be approved of by the *universal states*. In affairs of consequence, he is obliged to assemble the council of police, where he always presides, as well as in the *council of justice*, the vacating offices are *pro tempore* in the gift of the *governor* till further orders from the *directors;* the *governor* has the care of the safety of the colony, and issues the necessary orders for that end; but when is to be protected against inimical invasions, he must assemble the grand *court-martial*, consisting of the commander, all the *captains*, and as many members of the *council of police*, as there are *military officers* in the *court-martial;* the *governor* presides in this assembly, and proposes what *he thinks* necessary for the security of the colony. Lastly, the *governor* (by his instructions) is obliged to protect and promote the reformed religion in *Surinam*. The governor has a *secretary*, who is paid by the *directors*. The maintenance of all the *officers* is paid out of taxes imposed on all the *inhabitants*, the *directors* pay only the *governor's salary* with a *part* of the pay of the *soldiers*, and the maintenance of the *garrison*.

The *council* of police and *criminal justice* consists of ten counsellors, (including the commander of the *forts* and *troops*, who bears the title of *prime counsellor*,) with the *attorney-general* and a *secretary*: it has already been observed that the governor constantly presides in this *council*.

The *council of justice* consists of the *governor*, and six other persons, to whom a secretary is added; civil affairs are managed by this *council*. But an appeal lies to the *universal states;* the *governor* has here only one *vote*, except where a casting vote is wanted, *in these cases his opinion is decisive;* the counsellors of *police* and *justice* draw no *salaries;* the members are elected by the votes of all and every of the inhabitants for *a double number*, out of which the *governor* appoints according to his *pleasure;* they must declare by *a solemn oath* that they will observe and maintain (in every part) the *charter* granted by the *universal states* to the society and that in every other business they will *conform* themselves to the *orders* which from time to time they may receive from the *directors*.

At Paramaribo there is also a chamber of small affairs, and *a chamber for the management of the affairs of orphans and unpossessed legacies*. The first

§Sir William Temple.

commerce so naturally point themselves out, that it is unnecessary to say much on that head; I will however slightly touch on a very few of the principal prospects now apparently already open to us.

Our timber trade is certainly capable of being made incomparably greater than that of other countries together, as well in quantity as in variety: but care should be taken that this trade should be put under such regulations as would prevent a waste of timber, (such as has been too fatally practised in the north of America,)* and at the same time make our timber by a proper mode of manufacturing, answer the European markets, equally with those of the west-India islands, where the superior quality of West-Florida timber stands acknowledged without a rival. Nothing is wanting to effect this, but the erecting a number of sawing-mills on the Dutch model, and the procuring a regularity of demand for the manufactured boards, &c. As a proof that large quantities of timber may be brought down to

consists of seven *commissaries* and one *secretary*; the last of four *overseers of orphans*, who have a *clerk* and *book-keeper* in their service.

The *military*, maintained at *Surinam*, consists of four companes of foot; the governor is their *colonel*, and *captain* of the first company; the *commandant of the forts* commands the second company; this last named gentleman together with the *captains, lieutenants,* and *ensigns* compose the inferior *court-martial. The seven provinces have promised to pay the expence of one man in each company,* on the return of the *war-office,* for the protection of the colony of Surinam. The whole colony is divided into eight parts; according to which division, we find a similar number of companies of *militia*; each commanded by a captain, the two first companies are composed of the inhabitants of *Paramaribo*, the third of the ward of *Thorarica*; the fourth and fifth from upper and lower *Commawina*; the sixth of *Cottika* and *Perika*, the seventh from *Paulus-Creek*; and the eighth of the *Jewish nation.*

The churches, the clergy, and lecturers are maintained at the expence of the inhabitants; the clergy meet yearly in the month of February to consider the state and necessities of the church. In this assembly (known by the name of *Conventus Deputatorum*) a counsellor of the court of police presides, with the title of political commissary.

As an addendum, every man without exception, whether a native of the mother country or not, is obliged on his arrival to take the *oath of fealty* to this *noble society.*

I do not present my readers with the above account because I think a general monopoly is ever likely to take place, but to let them have an opportunity of comparing their own happy circumstances with those of a people where monopolists preside, considering at the same time that even these monopolizers through poverty are obliged to allow extraordinary privileges; I wish to put every person on his guard even against particular monopolies or monopolies in disguise.

All Dutch ships pay three guilders per last, both at entering and clearing out of the colony, which is about 2s. 9d. per ster, per ton, of every ship, port-charges, and the inhabitants pay a general poll-tax, black or white; each 50 lb. of sugar, and 2½ per cent, on all imported goods, and no vessels from the other Dutch colonies are allowed to enter here.

*For a small instance, dog-wood is used as fire-wood in New-York; it is a hard fine grained timber, fit for many uses to much better purposes.

our sea-side settlements, I beg leave to relate the following instance; *Mons. de la Gauterais*, formerly an officer in the service of his most Christian Majesty, but now residing as a planter on Pearl-river, some years ago had the command of the garrison of *Tombechbè*, during this tedious sequestration from christian inhabitants he thought of a scheme to turn this solitude to advantage, to effect this he employed several savages of the Chactaw nation in the cutting of cedar trees, and made those of his little garrison who were able and willing for a reward to do it, square the timber, and form some of it into a very strong frame of a house of about sixty foot by twenty-five, which frame was next spring erected on an island or gravel bank, about two miles below the fort, and during the summer season, and recess of the waters, this frame was filled with the squared timber, which was effectually secured against removing. When he had this raft completed, he got leave of absence, and watched the time of the rising waters, which at length took the raft away and him with four people upon it; this raft drew about 12 feet of water, and came down the river without let or hindrance, carrying before it every obstacle (even bending large trees under it) all the way down to Mobile, and by some neglect even as far down the bay as the present Fort Croftown when thus we see that an enormous raft containing upwards of five hundred tuns of solid timber may be with facility brought down the *Tombechbe*, above three hundred miles, what have we not a right to expect in the *Mississippi*.

Naval stores are likewise an article of immense speculation in both provinces; West-Florida already supplies Spain with considerable quantities; no province can so profitably furnish Madiera with corn and pipe-staves than West-Florida, and in return supply itself and other provinces with wines: the above named fisheries likewise may yield a very profitable commercial benefit. The trade for furs is a fifth article of importance, and is already very great; to mention *rice, indigo, tobacco,* &c. would be superfluous. But I have one material point to mention, I have often wondered at the stupidity of people let loose in a certain part of the field of commerce: I mean the trade with the Spaniards, the deceitful appearance of which has led many into ruin, and yet intoxicates every one who is so unfortunate as to stray within its bounds to a degree of madness. The generality of traders this way, consider only the price of things as they formerly stood, and not that these articles are much fallen in their price on the

Spanish coasts; the danger attending this trade they despise, though it too often proves fatal; the delays on the coast are not so much as thought of, and the perfidy of the Spanish traders forgot; the underselling by Dutch and French traders, although very real, is thought chimerical, and so on; in short, the prospect is here so gilded over, that nothing but profits appear, where nothing but certain and inevitable loss can be expected; the fatal mistake of overstocking every part of America with European manufacturers, is a kind of obligation, or rather compulsion on us to strike out into such pernicious branches of trade to the ruin of those who are interested, as well here as in Europe. Be then advised, and let no vain hopes intoxicate you, have patience, and if you import goods from Europe with views of profit, import them in moderate quantities, unless long credit or abundance of money enable you to wait patiently for the arrival of the Spaniards on our own coast; and let none of our vessels frequent their coasts except it be for the purpose of cutting fine timber and dye-wood, (and even these not often) or when you are very well secured by value left in your hands for the vessel to fetch a previously engaged, (but by no means by you) yet paid for cargo. Believe me, an American Spaniard will give you more for your goods at your own house than he will on his coast, and by coming to us he seems to loose all his cunning, which in fact is only owing to our eagerness of compleating a voyage when we arrive on their coasts. More branches of trade are not at present to be mentioned, because none have as yet appeared in this country to be worth pursuing, but an immense variety must and will in the course of nature make their appearance.

Having now explained my meaning in respect of the commercial interest of the country, I will say a few words on the benefits arising from the population of these and other similar provinces. It is an undoubted fact, which has long been long ago taken notice of,* that one man transplanted into the colonies creates work for four or five in the mother country; a late writer§ agrees with this in saying, that numbers of people who would have migrated to foreign countries, or done worse, are, by coming here preserved, and when become rich, return home. I must, however, differ in opinion from a third,† who

*By Sir Josiah Child, in his discourse on trade, p. 149.
§Dr. Campbell's considerations on the sugar trade.
†Political essays, p. 359.

has quoted both these, and expresses his opinion, by saying that emigrants ought to be made settle where they may be beneficial to Great-Britain; his saying would be just if a migration to the northern colonies was pernicious, but this is so far from being the case, that people there love room as much as any where; the continual and of late excessive importation of settlers into these colonies is an evident proof of this, for by means of these, even the natives are obliged to look out for habitations where they are more at liberty to encrease, not only the necessaries of life to which they are at present confined, but also profitable staples for the mother country. A very striking proof of this appears at present in every one of the northern colonies, though in none so much as in Connecticut, whose very hill-tops are inhabited full; those migrations ought therefore to be encouraged, because they will naturally find their way to the more favourable climates and soil of the southern colonies. It is also the interest of the southern colonies to encourage the migrations from the northward, for fear these last might by being confined within their prolific regions be in time induced to think of forming bodies of modern Goths and Vandals to overrun and invade the territories of their more happily situated southern neighbours. I am therefore induced to think, that (however strange it may appear that a people in one hundred and fifty years time should increase so rapidly as these have done) there is a necessity even now to enlarge their territories; in this I agree with the last quoted writer, and if Britain wants to secure the dependency of these colonies, it is absolutely incumbent on her to allow this, contrary to some late ill-judged schemes as this may appear. Can she suppose that those countries will remain unsettled? No, the prospect is too fine, numbers of families have through the natural course of the increase of mankind settled themselves on the lands intended lately for a new province on the Ohio: if I am rightly informed, no less than fifteen hundred families already, and many others on places not yet so much as thought of in Britain. These are the people who are most likely to form independencies, and not the known provinces; what ideas have these people of the paltry fear of a war with a few savages, (for in the northern America they are but few) not that I would be thought to imagine that these last mentioned provinces would not strike in with them in case of favorable opportunities; what if some enterprising genius among those dispersed inhabitants should form an idea of becoming a great man, and by insinuating himself among the most western tribes of savages (who never having had intercourse with us, nor any other white people, such for instance are the *Nadouessins,* who are said to be a people remarkably tractable) would incorporate a body of whites with them, and forming combinations with other neighboring nations, and thus erect themselves into a regular state. Such a state would undoubtedly soon become formidable, and be daily increased by such men as for want of room would leave the crouded provinces. These, supposing them at first vagabounds, or little better, when they should find themselves in a place where they might possess the sweets of life at ease, would find it their interest to become united with the common weal; therefore, I say again, that since for want of room, people to the number of at least five hundred thousand men, brought up to farming, are now without employ through lack of land, and therefore they are obliged to migrate,) and settle without the limits of the established governments, and thus it is highly incumbent on Britain to enlarge those limits, or form new provinces; a few paltry tribes of savages who retain the ancient grudge against us are no obstacle; the emigrants will soon bear down those melting remains of a people, who having lost their country, cannot fail to hate us, on that account, and in a kind of despair will rather choose to be destroyed than to incorporate with us. Such bodies as above described being once formed, can any man in his senses suppose, that possessed with the spirit and ideas of freedom, for which the Americans are so remarkable, they would after a course of thirty or forty years enjoyment of such an independency be brought to submit themselves to any imposed government? By no means; for not to mention their inaccessibility, men, who thus by a regular train of accidents become masters of a country by honest means, and improve it by robust industry, are not so easily dislodged as we may imagine history furnishes abundance of instances of such combinations of men, in a short time, becoming formidable; and even in General Oglethorpe's time a German with conceptions similar to those above named was found among the Creek savages, who, had he not engaged in his scheme entirely alone, and therefore by the Creeks, (although reluctantly) through the General's importunity and intrigues betrayed into his enemies hands, it would have been no wonder if such a man, with such ideas, joined to the amazing presence of mind and intrepidity he is said to have been possessed of, would have so nestled himself in the southern part of America, that we might have had no occasion to think of forming the two provinces of Florida in 1763; and we may venture to affirm, that had not the setlement of Georgia, just at that crisis become the object of Britain's attention, he would have gone on unnoticed till he had formed all the neighbouring nations into a regular governed body, too strong to be crushed by any power from abroad. But laying aside this reasoning, which I still insist upon is not only apparently probable, but may be looked upon as certainly to happen, unless the present bounds of the several provinces are enlarged; there are other cogent reasons to induce Britain to change her conduct in respect to this branch of her politicks, she ought to let the Americans spread themselves under regular governments over the continent, that they may be planters of staples, and thereby find the means of employing more hands in the mother country, by obtaining from her such of the necessaries of life as they find themselves obliged to have from Europe, at an easier rate than they could procure them among themselves. By suffering the upper latitude of the Mississippi, and the banks of the Ohio to be cultivated under the influence of a regular and civil government, we shall increase such plenty of necessaries in the mother country, that all connection with foreigners who have the balance of trade against Britain will become needless, any further than in selling them British manufactures for money, instead of what is now the case of sending bullion, besides the stated quota of manufactures, in exchange for hemp, iron, flax, and other such necessary and bulky articles. When we consider, that during the period of time in which these higher countries enjoyed and were protected by a state of regular government under the French, one winter (so long ago as twenty-eight years) furnished new Orleans with eight hundred thousand weight of good flour from Illinois*, we must be sensible at what a reasonable rate this and other necessaries of life could be procured at New Orleans and Pensacola, and from thence to East-Florida and the West-Indies, where the Pennsylvanians and New-Yorkers at present insist on their own price, in so much that flour now is commonly at from 20$s.$ to 22$s.$ per cent their currency at home; whereas I remember that even in 1760, during the war, 16$s.$ and 18$s.$ was thought a high price. This alone demonstrates the necessity of such a measure; would not such a fresh supply spread its influence even across the Atlantic to the poor in Britain? no doubt it would,

*Du Pratz.

the navigation down these rivers is nothing in difficulty or expense compared with that of the European rivers, from whence the materials come for the supply of Britain. The inhabitants of the upper counties of Virginia and Pennsylvania would be obliged to drop the bulky commodities they now raise, if they saw the inhabitants of upper Mississippi, the Ohio, &c. raise the same with redoubled ardor, and in a perhaps four-fold quantity, at the same time bringing them to market at much less than half the expense themselves can do it at, and instead of the said coarse, heavy, and bulky articles, they would necessarily be induced to think seriously of a vigorous culture of wine, silk, oil, and similar necessary, light, and valuable products, able to bear the heavy tax of land carriage to which those countries are by nature subjected. In case of a rupture with any power who should think proper to attack the southern provinces, especially West-Florida, would they not most naturally be supplied with provisions, ammunition, and soldiers by way of the river from those settlements in the higher latitudes? nor is it on account of hemp and flax, &c. alone, which may be said to be still objects of speculation, we see the tobacco culture which is a branch of vast importance to the revenue as it were daily mouldering away into its primitive non-existence, and therefore new lands for this culture ought to be found and improved; the want of this supply of the revenue will soon be felt in the treasury, and Britain will find too late that she would have done well to have gone not only to he expense of the establishment of these new provinces, but even to the expence of that tremendous *bug-bear*, a *war*, with the remains of the ancient inimical tribes of savages; unless indeed, she intended the proclamation of October 7, 1763, as a meer matter of form and wished the back settlements quietly to spread themselves without any expence to the mother-country; but should this be the case, the mistakes of such policy I think are fully refuted in the foregoing pages; The spreading of people over the continent will make them produce great quantities of staples, and Britain hereby would secure a long period for the dependence of the colonies upon her, while she would have nothing to do but provide markets for the produce of the labour of her sons. All Europe is sensible of this; every writer I meet with is either of my opinion, or, absurdly and continually contradicts and confutes his own maintenance of the contrary in his own writings. I beg leave, to quote a very sensible Frenchman (whose name I know not) on this subject, "it is true," says he, "that the free and happy situation of North-America, may much disorder our European systems,* particularly, if the English colonies attain to the not paying imposed duties. Vast, fertile, and new tracts, where neither customs, taxes, nor military troops, will take place, because they need fear no invasion, will deserve the most serious attention from all our governments; and policy will be obliged more than ever to turn all her views on the side of beneficence. Those states which will be latest in using this reflection, will find themselves utterly incapable of mending the evil; for the greatness of punishments or weight of slavery only serve to increase ill-blood." This is not the way to mend the matter, and I think certainly that no remedy will possibly be found out against such a number of evils, as much necessarily attend the keeping the colonies confined to their present bounds, whereas, by only allowing people to spread over the country, they will all be effectually prevented.

To give my reader an idea of what room there is yet left for people to plant themselves, I must tell him, that between the latitudes of 31 to 46, which I judge to be the most habitable country from the Atlantic to the Pacific ocean, and which cannot fail by the course of nature to become the country of those Americans who were originally planted from England, Holland, and France, and emigrated from Germany, on, and to the shores which bound the Atlantic westward, and who now generally speak English; within these limits I say are contained no less than two thousand one hundred and thirty-five millions, and forty thousand square acres, exclusive of the known waters; at least twice as much as the vast empire of *China* is said to contain, and allowing each family to possess fifty acres of land, there is room for forty two millions, seven hundred and eight families to live here.

To conclude this discourse on the settling of new provinces, I will just observe that the daily emigrations of the people beyond the frontiers, even at the risque of the dreadful situation of being exposed to savage incursions, is the most evident proof of the necessity of this measure to the at present crouded Americans, if not to Britain. Likewise it is a proof how little the present Americans who inhabit the frontier countries value so insignificant a foe as the re-

---

*The French word is *Combinaisons Europeanes*, which may mean European alliances, combinations, or junctions, but I cannot find English sense in using either, therefore I suppose he means systems, or he may have meant European plots which the word also bears.

mains of the ancient inimical tribes of savages, or even the precautions in law of super-intendants, their delegates, and other formerly useful officers.

To say a word or two about the political history of these provinces, may be thought incumbent on me, but little can be said of moment on that subject as yet; except that Major Ogilvie in taking possession of the eastern province, by his impolitick behaviour caused all the Spaniards to remove to Havanna, which was a deadly wound to the province, never to be cured again, notwithstanding the inviting means used by Governor Grant, who succeeded him, to retain the remainder, and to make them all return;—that Governor Grant used all possible means to encourage indigo and rice plantings;—that the crown would not allow the transfer of Spanish landed interest to be good, although mentioned in the articles of peace;—that the said governor reigned supreme without controul, and in peace, notwithstanding the frequent murmurs of the people, and the presentments of the grand juries, occasioned by his not calling an assembly, which they thought was a duty incumbent on him;—there was also a complaint of the contingent money of five thousand pounds per annum, for seven years, not being so very visibly expended on highways, bridges, ferries, and such other necessary things as the people could have wished; finally, Governor Grant departed the province for Britain in 1772, on account of his bad state of health, having a pompous address presented to him by his council, which was re-echoed by a few of the inhabitants; —that he was succeeded by Major Moultrie as lieutenant-governor, who was again succeeded by Patrick Tonyn, Esq; as governor in chief, who enjoys the command there now;—that in the northern extremity of the province at the mouth of St. Mary's river, a place is now founded, which, to all appearance, by its situation and superior entrance into its harbour compared with St. Augustine, will draw the seat of government away from the latter;—that the cultivation of many articles, especially indigo, begins to wear a pleasing and promising aspect.

Concerning West-Florida, I can only say, that discord has early made her appearance in the counsels of this province;—that disaffection to Governor Johnson has been the occasion of its being so late as it was in rearing its head;—that Montfort Brown, Esq; succeeded him as commander in chief in quality of lieutenant-governor, who was succeeded by .... Elliot, Esq; who unhappily ended his days; when the lieutenant-governor again succeeded him, and was a second time replaced by Governor Chester, who is universally esteemed, and under whose auspices the province is in a thriving condition.

About March 1772, the freeholders being met for the purpose of electing their representatives, but finding that the writs mentioned a continuance of the assembly for three years, (whereas they had heretofore held annual elections) they now added this condition to their votes, that the new elected members were to continue only one year; the mode of return on the writ the governor took ill, and refused to accept it, insisting that the members should be returned for three years, but all in vain, the freeholders remained inflexible, and rather than not have annual elections, they chose to remain without representatives which has been the case ever since. The settlement of this province goes on very rapidly, and its different valuable products increase a-pace.

I shall now treat of a subject in which I confess myself not to have so much skill as I could wish; I speak of it from such experience as my temper, inquisitive into the mysteries of nature, has furnished me with during my stay in this country; I mean the diseases most frequent in those provinces. But by way of prologue, I shall enquire a little into the universally dreaded, though chimerical unhealthiness of this climate. Dr. Lind* enumerates some proofs or signs of an unhealthy country, which are as follows, viz, 1st. Sudden and great alterations in the air, from intolerable heat, and chilling cold; this is perceived as soon as the sun is set, and for the most part is accompanied with a very heavy dew, and shews an unhealthy swampy soil; this is perfectly the case on St. John's river, and about Nassau-river, in East-Florida, likewise at Mobile, and Campbellton, in West-Florida. At Pensacola, and from thence East: there is little or none of this perceivable; at Orleans, and on the Mississippi I was not sensible of any sudden alterations of this kind: and on enquiry found it not to be generally complained of. Neither of those provinces are so subject to this unhealthy variability as Georgia and Carolina, especially the last; however, I do not find that any person need be much under uneasiness about this any where, not even a newly arrived European; the chief care necessary to such at seasons when these sudden alterations take place, is to avoid being exposed to the night air, and after

*2d edition p. 1. ch. 4. sect. 2, p. 137, et seq. London 1771.

sun set, to add some more clothes to those worn in the day time; and if on a journey in the woods, never forget keeping a very large fire at night in order to rarify this dangerous air. His second sign is, thick noisome fogs, arising chiefly after sun-set, from the vallies, and more particularly from the mud, slime, and other impurities; this I never perceived to a great degree in any part of these provinces; St. John's alone is very subject to thick nasty fogs of all kinds, the sea-side where soft salt marshes are frequent, is often troubled with these, and these marshes a little before rain emit a most horrid, and to me a suffocating stench; the northern part of East-Florida is very full of these kind of marshes, in Georgia and Carolina they cover a prodigious surface; the marshes in the south east of East-Florida are of a different nature, and in West-Florida I remember very few of this kind; I have, however, never heard any thing said contrary to a prevailing opinion of great salubrity in dwelling near the sea side, even among the thickest of these marshes; nor have I ever heard any notice taken of the above stench by the inhabitants, except as being a certain indication of rain; where this kind of marshes are situate in brackish water, the situation is beyond doubt very unhealthy, the wan complexion and miserable mien of the generality of the inhabitants of such districts too plainly evinces it—We know by experience, that all such fenny countries on every part of the earth labour under the same unhealthy calamity, but the comfort is, their areas are every where very small when compared with the more **salutary situations** which abound almost universally in every province, and in every climate.

The third is, uncommon swarms of flies, gnats, and other insects, which attend putrid air, and unhealthy places covered with wood; the first of these is not common here, except at indigo works; and if these, according to my former caution* be a little remote from the dwelling, I never saw nor heard of any bad consequences with regard to health attending them. I have already mentioned the bad effects of these voracious vermin on cattle of all kind at an indigo work; gnats, here called musquitos, are vastly numerous in some spots, but after we have passed some distance the brackish waters in every river, they diminish, and at last we find none; the Mississippi however is an exception, but on this river they are not in such plenty at the freshes as below, at the *rigolets*, on pearl river, and at the *Riviere*

---
*Page 139.

aux Boeufs, and on Dolphin-island, likewise in Santa-Rosa bay, and in the bays of St. Andrews, St. Josephs, and at St. George's sound and islands, they are intollerable, and all the sea coast of both provinces is exceedingly pestered with them. I travelled across the peninsula in 1769, in the months of June, July, August, and September, but found none at all in the interior part; I never heard these being attended by any fatal disease except their troublesome bites, which sometimes cause inflammations, specially in the legs; but the inhabitants of the western province so effectualy fortify themselves against those vermin with musqueto nets, tents, and *Baires*,† that whether at home or travelling, they are not in the least danger from the attacks of this terrible and bold, though diminutive enemy, whose destruction gradually takes place as the woods daily diminish; in very dry hot summers, scarce any of these vermin are seen; a very dry hot air causing the deaths of numberless *animaculae* of every kind, their effluvia, even of those that are imperceptible to the naked eye, arising or exhaling from ponds, marshes, swamps, &c. must spread a great quantity of noxious vapours through the atmosphere,* and consequently corrupt the air, and spread disease throughout their vicinity; this misfortune will likewise cease on opening the country, till then let me advise every new comer, particularly a person of a gross habit of body, to be careful of his constitution, a wine-bibber, or rum guzzler, with such a plethoric habit, can hardly avoid falling a prey to this bad air.

His fourth, is the quick corruption of butcher's meat, &c. this, if the case at all in any of these provinces, is not common, I remember indeed to have heard a complaint of this happening at Pensacola in 1765, which was a sickly season, but in all my journies through these provinces I never experienced it, on the contrary I could always by some means or other preserve my venison, or beef, when there was a necessity for doing it.

The fifth reason, says the Doctor, is a sort of sandy soil, among others he says, such as that at Pensacola; I beg leave to inform the Doctor, that the sand about Pensacola, and throughout these provinces, is a coarse, gritty and gravelly sand of various colours, though chiefly red and white, that on this sandy soil many excellent salubrious

---

†*Baires* are a kind of tent made of light coarse cloth, like canvas gauze, called by the French *villemontiers*.
*See page 10 of this volume.

herbs grow, which serve as food to innumerable herds of cattle, and when cleared they are improvable by culture; this is not the case with the hot sandy desarts of South-America, Africa, and Asia, over which the *samiel* wind passes, nor do I believe that the oldest person in Florida remembers any sudden, hot and suffocating gusts, or blasts, from which he has ever been obliged to turn his face in order to draw breath; there are no open plains of sand in North-America, they are all covered with trees rooted in the lower strata of these sandy tracts; the only plain of sand I know without trees, is at, or near the head of St. Lucia, in East-Florida, which however is not extensive, and to it numerous herds of roe-deer resort during night from the adjacent woods; I think, therefore, that neither of the Floridas have any ill consequences to dread from their land winds; moreover, if we consider, that the land-winds come from the west and north-west; geography will tell us, that they range over innumerable acres of oak land, consequently clay ground,* and that only the sea coasts, and from twenty to about an hundred miles off, are any ways sandy; which being so constantly fanned by the wholesome sea-breezes, (so remarkable in these provinces) could not have any fatal effects, even if the ground was of the nature of the Lybian desarts; thus we see again how men reasoning from mere theory, are liable to commit mistakes.

To treat of the diseases to which the human frame is most liable here, in the same regular manner, as I believe I have done of every preceeding article, I will divide them into two classes, viz. acute and chronic, and the first again into two orders, viz, those of the summer, and those of the winter.

Fevers are the first of the summer diseases; the ancients have ages ago made an observation, that the season of the reign of this terrible disorder was always preceeded by an atmosphere laden with great heats, and much rain, for some time; the modern writers seem to me to be generally of the same opinion; this is exactly the case in all the southern provinces; for fevers begin to take place in some districts more, in some less, about the latter end of July, and in August, and continue throughout September, and part of October, just the season immediately succeeding our greatest rains; and most

---

*Excepting the few hommocks near the sea, which are oak land, but most of them sand.

violent heats; here I will notice a* remark which I have read long ago, and I find it confirmed in all climates, "That the middle of the third month was observed to be the period of the greatest rage of epidemical disorders." Those districts which lay near to low rice fields, particularly in back swamps) and to such indigo works, where the planter is obliged to make reservoirs of water, are most liable to these disorders, after, and during the latter part of an excessive drought; because in those neighbourhoods the air is at such times most prodigiously loaden with corrupt moist effluvia; for this same reason, cool rainy summers will make those places more healthy than dryer spots, because during such a season all the above mentioned noxious exhalations do not take place in so great a degree, and the air is kept cool by the frequency of the showers; however such situations will never be so common in the Floridas as in Carolina or Georgia, the quantity of good wholesome fresh running water being infinitely greater, and consequently little necessity of making stagnating ponds or dams.

It must be allowed, that all fevers however dissimilar in appearance, proceed from the same origin; nature only works with more or less violence to rid herself of what is detrimental to her.

The *Ephemera*, or day fever, occasioned by a meer increase of the velocity of the blood, by means of a fit of drunkenness, or debauch, or originating from violent exercise during the heat of the day is too frequently seen here; but as it is seldom of a longer continuance than eighteen or twenty hours, it has not often dangerous consequences, and may be avoided by every person; I shall content myself with barely observing, that some cooling acidulated liquid aliments will soon abate its violence, bleeding may likewise be of use to restrain its force.

The continual fever, or inflammatory fever, is sometimes, though rarely experienced in this climate, but seldom attended by those dreadful symptoms and fatality, which accompany the same kind of fever, though of a more violent class in the countries immediately between the tropics: this, in its common form lasts about ten or twelve days, beginning to abate its violence in general after the seventh; the fourth or fifth is often fatal. I am persuaded, that whenever the yellow fever has made its appearance in the Floridas,

---
*Inquiry concerning the cause of the pestilence, and the diseases in fleets and armies. London 1759.

it was imported from Jamaica or Havannah, as was the case in 1765, which (by the way) was almost universally an unhealthy *Aera,* as well in Europe as elsewhere. This continual fever begins with an excessive heat of the whole body, continued, though not violent head ach, great drought of the tongue and palate, and consequently a continual desire to drink; those people who die of this disorder, generally depart on the fourth day, and I am of opinion, that few are carried off by it, except such as are kept too close confined from the fresh air: I would recommend the keeping the sun out of the room, but to admit as much air as will gently ventilate it; a cooling diet, such as rice gruel, barley water, infusions of baum, or sage, and lemonade, which is lime juice, water, and very little sugar; lime juice, syrup of lemons, and currant jelly should moderately enter into every part of the patient's diet; avoid all salt, spices, spiritous liquors or generous wines; a gentle purge of glauber salt, with a few grains of *kermes mineral,* and some drops of oil of mint is generally given on the first appearance of the disease; the effects of this are forwarded by frequent draughts of warm chicken broth. During the operation of this, avoid all acids: bleeding (especially if the disease makes a violent attack, and the patient is of a plethoric habit) is indispensably necessary, the patient ought by all means to avoid motion, and notwithstanding the above caution of admitting air in the room, keep himself covered, and be careful not to throw his bed cloths a-side. If the symptoms abate after the above mixture, emetics are commonly subscribed; if it still continues, particularly if attended with *delirium,* lethargic symptoms, or their reverse, blisters are applied; and in great watchfulness some *laudanum* is used; if worms are suspected, an infusion of Indian[†] pink root, (a very common plant here) leaves, wood, and all, is made use of as tea; but this plant possessing a pretty strong narcotic quality, ought to be used with caution: in excessive heats some grains of *sal nitri* are added to the liquors administered to the patient, and as soon as the fever begins to abate, some orange, lime or lemon juice, saturated with salt of wormwood, is given by a small tea-cup full every two hours.

The yellow fever being sometimes imported here, it may be necessary to describe this *Proteus* among diseases, which I have frequently seen, and which I myself have suffered in Jamaica under one of its various forms. I chuse to follow the description of *Dr. Rouppe,* in his

---

†Lonicoera.

de morb. navigant. as quoted by *Dr. Lind,* because I find no other author who has done it so exactly, and if I had not seen his account I should not perhaps have recollected half the symptoms he mentions, altho' I now perfectly remember to have seen that dire distemper in every shape he paints it in. He seems to speak of it as an illness to which seamen only are subject in that climate (*Curacoa;*) he may have been induced to think so by the disorder not raging on shore; the florid complexion of most of the people in the island sufficiently shewing it to be a very healthy spot; but I find by his description, that it was the real yellow fever contracted at *St. Eustatius*; he calls it the putrid colliquative or spotted fever; it raged in the Dutch ship of war Princess Carolina, on board of which this physician arrived at *St. Eustatuis* on the 1st of August, 1760, on the 11th they sailed from thence for *Curacoa,* which harbour they entered on the 19th, and then (whether he means on the passage, or the very day he arrived, seems doubtful) twenty people were sick, some of these had head-aches without fevers, and some were afflicted with a true bilious cholic; but they were by an easy cure again restored to the enjoyment of their former health: the sense of his account of the disorder, I take to be as follows, viz.

"In the beginning of our stay in the isle of *Curacoa,* as before mentioned, the diseases which occurred most frequently were headaches and bilious cholics, which were soon cured; this changed into true choleric complaints, gradually increasing with pain much more dangerous than the former, wonderfully tormenting the patients; the disorder begun with a very great burning heat about the *praecordia* (sides of the upper part of the belly) griping stools, great anxiety and uneasiness, these were followed by bilious stools and vomitings, with great loss of strength, many of them being bedewed with abundance of cold sweat; if this continued, especially if a fever came on, (which was the manner in which it seized some with a high pulse, which continued in most about ten hours), then the lips began to swell, and a ghastly paleness seized the face; afterwards, the fever abating, they vomited abundance of dark coloured blood; at this period they generally died, or within a few hours after the appearance of these symptoms. Those who voided the above named and some blacker matter by stool, emitted a terrible offensive smell; but some of these were with difficulty cured; in the same manner it happened to those, who suffering of the fever, had, however no evacuation; others were only seized with a common bilious fever, and as much as I could judge, the major part of those were young men, or middle aged, robust, and before sickness the briskest and most cheerful; they had a heat about the *praecordium*, they vomited bile, or were making attempts to vomit and had an unquenchable thirst; some of these frequently found themselves cold, and by turns heats would seize them; then succeeded a hot itching of the whole body, with a high, full, and quick pulse; the tongue was yellowish or whitish—often encompassed with a green border on the edges, and always wet or moist.

"The disease continued in some to the second, in others to the third day; then the heat would abate of itself, and the natural pulse returned suddenly or unawares, which by little and little would sink, and at length become small and tremulous, in some there appeared *petechiae* about the breast, arms, and inside of the thighs; in others, large livid spots appeared; the strength of some was by all this so exceedingly exhausted, that on the least motion, the patient would swoon; besides an abundant sweat would arise on the whole body; the patient moreover would be anxious, fretful, uneasy, slightly delirious, very inattentive, valuing nothing, complaining of nothing, eluding questions, and yet at the same time almost always answering pertinently to them; in some, upon the declining of the pulse, a fiery heat would arise about the stomach, the lips swelling a little, the face becoming ghastly, shortly after they vomited discoloured matter, and at length died; others would be consumed by heat and griping stools, and discharged corrupt, stinking, and almost black blood by stool; some in the third, and others in the fourth day, would begin to acquire a yellow tinge in the white of the eye, and on the skin, which was an evil sign; moreover, the tongue would from day to day become whiter, and at length tremulous; they would always lay on their backs, thus as the disease increased, sometimes on the second, third, or at latest on the fourth day, an easy, calm death would follow.

"Blood drawn from the veins, in the heat of the fever, was bright red: it concreted and separated a yellow *serum*, just as in Europe. Those, who by meer dint of strength resisted the disease, and reached the fifth or seventh day, would have the whole body almost covered with small boils, or little painful red pimples, very difficult to be brought to a suppuration, resembling small pox of the confluent kind.

"At length, most of the diseased, particularly those who had reached beyond thirty years of age, and were of a bad habit of body, when they were seized with the disorder were overwhelmed with pain and heat about the stomach, with a continual retching, but yet bringing up little or nothing; in some the pulse increased for some hours, and again appeared in a short time to be in its natural state; it became low, the skin possessed its natural heat, the tongue was moist and white; on the first day of the disease a copious sweat would break out over the whole body, but no spots appeared, those whose sweat was little or none, had copious black and very foetid discharges, were troubled with gripings, and would often faint away suddenly; when the evacuations were trifling, ceased, or otherways remarkably lessened, and but little sweat pervaded; then the patients suffered greatly and were very restless; on the contrary, if it broke forth plentifully, they found themselves much better.

"Lastly, in all it stages from beginning to end they were all afflicted with constant watchings.

"A worthy lad about eighteen years old, who found himself well in the morning early, was, about 10 o'clock of the same morning taken with a head ach, and other feverish symptoms, he had a high full and quick pulse, he obstinately refused being blooded: the second day at evening he voided abundance of dark discoloured blood by vomit, and the third he died. Another of about sixteen was well in the evening, but we found him next morning ill, and utterly deprived of all sense, I examined his body, which was swelled and overspread with livid spots, his pulse almost entirely gone; some black blood of a sweetish taste flowed from his left ear and out of his nostrils, which continued to ooze out some hours after his disease, the corpse in a short time became wholly discoloured and livid, emitting a very disagreeable smell."

The French call this disorder *mal de Siam*, supposing it originally imported from thence into the islands; the Spaniards *vomito preto*, or black vomit; the Dutch *geele koorts*, which last conveys the same idea as the English yellow fever.

In general, when fevers are violent, the practice which prevails at present, is to have recourse to antimonial medicines, and as soon as a remission is brought about: the bark is administered in large doses.

Intermittents are endemial in all low situations, thus we see in all the provinces to the southward, particular places remarkable for a continuance of this disorder in them, such as more especially Jacksonburg, in South-Carolina, Savannah, in Georgia, Rolles-Town, and most of the settlements on St. John's, in East-Florida, at Campbelltown, near the mouth of the Escambe and at Mobile in West-Florida; this disease attacks people much in the same form as the continued fever, the first fit frequently lasting three days without intermission; physicians treat it nearly in the same manner at the last, but I have observed, that they are very averse to taking blood from a patient afflicted with this disorder, saying, that bleeding is a sure way to prolong the disease, although sometimes a small matter of blood is taken from people of a very gross habit of body, when the returning fits seemed to continue longer in point of time than at the first, the same diet is observed as in the continued fever, except when the patient is very weak, when strong broths well separated from the fat are frequently given; if delirious or comatose symptoms with pains in the back, &c. make their appearance, cooling medicines are used, during the paroxysms, Doctor James' powder or other antimonials, and on intermission the bark in copious doses is administered with success, and in obstinate head-aches recourse is had to blisters.

This is a very tedious disease, and whoever is afflicted with it should not too soon judge himself cured, but continue taking a bitter infusion, composed of the bark of the root of the *magnolia major* (which the French on the Mississippi substitute in lieu of Jesuit's bark) with Virginia heart, snake-root, rue, *sal absynth*: and pink root, in good Madeira or Lisbon wine.

People in general, suppose them even obliged to remain on the sickly spot during the fatal season, which is autumn, may by care, in a great measure shun this tedious illness, such as living on a more generous diet, especially animal food high seasoned, and a moderate glass of wine; avoiding a too great exposure to the then frequent sudden changes of air. They ought to use the cold bath often, wear garlic and camphire in the pockets, not expose themselves to rain, and above all keep warm and dry feet, and if got wet by rain not to change their clothes too suddenly; never go out of a morning fasting, but before you go to work, business, &c. eat a piece of bread, and drink a glass of the bitter infusion; avoid the night air, and keep some fire in the house, particularly in the mornings and evenings to rarify the damp air in the rooms, especially in the bed rooms which ought never to be on a lower floor, and should be in the eastern parts of the building exposed to the morning sun: by observing these rules the constitution of the human body will be less disposed to receive the impressions of a bad air.

An excellent thing to be given the negroes on a plantation before they go to work, is a wine glass full of the above bitter ingredients, and garlic infused in rum; and they should be encouraged to chew and smoke tobacco.

When a person is seized with a fit of the ague, he ought by no means to delay going to bed, and drink a draught of lime juice, and powder of chalk, while it is fermenting in the glass; this will bring on a sweat, and shorten the fit, or in the hot fit use some opiate if the patient is not delirious, this ought to be done as often as the paroxysms return.

The nervous fever, likewise called the slow fever, is known by a small, quick and low pulse, and by not affecting the patient with such violent heats as the others fevers, but with greater oppression about the *praecordium;* it does not make them so thirsty; the tongue is at first unusually moist, and looks white, though at last it becomes dry, and looks brown or inflamed; continual heats are felt in the palms of the hands, heats and chills return alternately very quick, a copious clammy weakening sweat, excessive lowness of spirits, restlessness, being drowsy without power of sleeping, pain and giddiness of the head, ringing in the ears, and if it lasts long, the tendons are often affected with a sort of cramp; deafness, deliriums, continual lethargic fits, insensibility and stupor are the constant attendants of this disorder when in its last stages.

This is a most treacherous disorder, and by affecting the sufferer with only slight symptoms of weariness and weakness, attended with frequent yawning and stretchings, a slight giddiness and loss of appeite, and a great heat in the forehead, makes people neglect an early application to the physican, and thus they endanger themselves much, though in people of a robust constitution who are much exposed to the sun, it will often appear for the first day or two with violent symptoms; this fever will last sometimes for twenty days or more without any apparent abatement; it generally attacks people who have been exposed to unusual fatigue, or such as are naturally of a weak constitution. Vomits are the remedy to which recourse is most usually had in this disorder.

Physicians steadily, and almost totally avoid bleeding and purging, till after a free use of the *Ipecacuana*, and even then their cathartic prescriptions are rarely any other than manna and salts, and after the gentle purges obtained by this method, they order a free use of rich chicken broth, and the above described juice of lemons saturated with *sal absynth:* this they generally continue until the disease changes into an intermittent fever, and then treat it in the manner last mentioned; frequently also applying blisters.

This fever more particularly than any other disorder, bears hardest on the patient towards sunset. The diet commonly prescribed is sago, chicken broth and panado, with some small matter of wine and loaf sugar in the first and last; infusions of sage and baum, together with wine whey, are the drink mostly thought proper during the continuance of this disorder.

The use of bark is generally blamed as productive of dropsy; jaundice, the ague cake,* and other inveterate chronical disorders, but it is certain that the bark is blameless; it is the fault of the physicians, who too late and with too much caution use this blessed remedy, which seems as purposely designed by Providence to relieve us in those tedious disorders; or it is that of the patient himself, who, prejudiced against this excellent remedy, refuses to take it till it is too late, and thereby brings upon himself the above diseases, which are consequences of the fevers, not of this great specific.

The above mentioned fevers and unusal hardships in travelling, &c. as well as excesses at plentiful tables full of variety, often bring on a severe bloody flux especially in autumn, and if this makes its appearance with hard dry and bloody stools, the disease is dangerous; brisk purges and clysters of Castile soap, and some of the hot seeds are used to expell these; when the desired end is obtained, gentle emetics are called in; I have known people find great relief from a decoction of logwood and pomgranate skins, others again it would not help in the least; a new honey comb inclosed in an apple scooped out, and then roasted before the fire, has often proved a speedy and very effectual remedy; calcined hartshorn, a nauseous medicine, is nonsense; bark of *summach* is a good medicine, but there being a dangerous kind, it ought to be gathered by a skillful hand; the bark of the *liquidambar styraciflua; aceris folio* mentioned in page 20, to-

---

*A hardness in the region of the spleen, one of the consequences of long continued fevers, and by the Dutch Creoles distinguished by this name, they call it in their own language *koek in de buyck*, or simply *de koek*.

gether with the gum exuding from the same tree, is generally found efficacious; a wine glass full of the juice of lemons mixed with some common salt has often proved a most excellent, safe, and general specific; the frequent chewing of cinnamon and camomile flowers, especially when a weak stomach vexes the patient, has a noble effect; avoid all vegetable food except rice, eat roasted rather than boiled flesh; salted beef need not to be avoided; use often veal, jellies and salop; use a great deal of mustard; I know by experience of many as well as my own, that Dr. Barry's observation of vegetables not being so easily assimilated as animal food, is in the strictest sense universally true, and in an obstinate continuance of this disorder, vegetables even our common wheaten bread are not at all digested, but most generally pass through the body unaltered. When this disease changes into a chronic habitual flux, it will be necessary to use pills of equal parts of rhubarb and *ipecacuana*, mixed with some liquid opiate, and use weak lime water for common drink: if this does not prove a specific, let the patient be removed to some other clime, for no remedy will affect that disorder in the same climate, where it was originally contracted, claret or port ought to be their constant drink in this disease, and spirituous liquors ought by all means to be avoided; rum, a cursed bane of health and of society; is too often and indiscriminately applied to every disease as an universal *arcaum*.

The *cholera morbus* is likewise a consequence of intemperate meals, and when it is not occasioned by any food peculiarly repugnant to the stomach it often proves fatal.

Debauch of every kind, particularly unseasonable sitting up, is most frequently productive of some of the most dreadful disorders, and excessive passions of the mind sometimes produce the same effects.

Excess in venery is generally productive of the most violent and obstinate disorders, principally inflammatory fevers, and obstinate fluxes.

There is a disease which the French call *La Tytanose*, which affects people in the western parts of Florida, and will attack them with prodigious violence upon being wounded even in the slightest manner: if during the hot months a splinter be run into the flesh, the patients are attacked with violent contorsive spasms, and generally die in about eighteen or twenty hours.

I never saw any person afflicted with this dreadful disorder, but from the similarity of the name with the latin *Tetanus*, and from by being told, that opium and camphire are much used to procure relief, I take it to be the locked jaw, with which I saw a young man die at Mobile, Mr. Lind* recommends copious external applications of opium, and the cold bath; and gives some imperfect account of mercurial ointment having lately proved an efficacious remedy; the hint was perhaps necessary to be inserted here.

Angina Suffocativa, or the putrid sore-throat, sometimes appears here, this is a contagious distemper, and rages in America mostly among the youth; it generally begins with a slow fever attended with great lassitude and a low pulse; this is succeeded by a sore throat with white spots near the *uvula*, and if it be not immediately taken notice of, the patient soon becomes past hopes, and generally dies within 24 hours after the first severe attack of the fever; the physicians in Carolina and Georgia prescribe first mercurial purges, and order a gargle of borax, dragon's blood and Armenian bole in vinegar and honey; and the throat is anointed frequently by help of a feather, with a mixture of balsam of sulphur, tincture of myrh, honey, loaf sugar, and yolk of eggs; the principal part of the cure is to attend the disease early, the least neglect being dangerous.

The dry belly-ach is a very painful and tormenting disorder, though rarely fatal; it is occasioned by cold damp lodgings, and being exposed to the night air; but most frequently in all climates by an excessive use of the vegetable acid juices, which are all extremely astringent in their nature: and when this disorder proceeds from too liberal an use of punch, rheumatic pains and paralytic affections of the nerves are its constant consequences and attendants, with loss of the proper use of limbs: (often for life) the most usual symptoms are the vomiting of bile, with the most obstinate costive habit imaginable; and when stools are procured, the excrements are excessive hard, and in round balls like horse dung: all this is attended with the most excruciating pain in the bowels, and a clammy sweat the method of cure is by administering emetics of the antimonial kind, which often also procure a stool; this is the only thing that can relieve the poor sufferers; the warm steam of hot herb baths, clysters of the *tinct. Thebaic.* in lukewarm milk, and emollient plaisters in which opium enters, applied to the stomach and belly; bitter purging salts

---

*Hot Climates, second edition, page 285-286.

and manna, and infusions of sena leaves, after the middle vein is stript out of them; in the severe attacks of the pain, opiates are used, and too often that cursed *arcanam* of the vulgar among the English; I mean rum and other distilled spirits, which in this disorder too often prove fatal poisons; the oil of *Palmae Christi*, by three or four spoonfuls has sometimes proved effectual; oil of almonds and of olives, having been given with success; after all medicines have failed, I once applied to a mulato woman, who was a noted empyric in the island of Curacoa, where I was attacked by this distemper: she ordered a clyster of sweet milk, tobacco and brown sugar, which gave some slight relief, but after a while the painful symptoms of the disease seemed to be as excruciating as ever; she then gathered some handfuls of the leaves of a shrub which is there called *Wild Carpat;** these she boiled like spinnage, and made them into seven or eight balls of the size of walnuts, put them in a plate, and poured oil olive on them, and a little pepper; this kind of salad she made me eat with a piece of bread, when I observed to her, that she ought to have added vinegar to have made a perfect sallad; she answered, that vinegar in my case was poison; in half an hour after the use of this mess, a stool (the first in twenty-three days) was procured, which was followed by five or six more that very afternoon; and she then gave me for some days an intensely bitter mixture, in which I perceived the juice of aloes predominant, but could not learn the composition. This kept me in a lax habit of body, and in about fourteen days I was enabled to pursue my ordinary avocations; camphire and opium enter into all the purgative prescriptions I have seen ordered in this disease, by the physicians of the south.

There is an instantaneous fatal disorder which the French call *un coup de Soleil*, i.e. literally, a stroke of the sun; of this I remember one instance during my stay in West-Florida, when it killed a child of about twelve years old on the spot between the hours of eleven and twelve in the forenoon, the time, as I am informed, in which it always takes place; by instantly applying cold water to the crown of the head, I am told, its fatality is prevented; likewise by cupping the crown of the head: what its symptoms are, I have not seen, but by the descriptions, I take it to be a fever, which so violently attacks the patient, that it causes instant death. This disorder occurs

---

*Carpat, is the Dutch name of *Palma Christi*; I have long in vain tried to find the shrub, from which these leaves were taken; all I know of them is, they are *Lanceolate*, of a lucid green, and boil very tender.

very seldom, and as it is so very easily guarded against, persons who are attacked by it, are in a great measure blameable for their own misfortune, particularly if they know the country; the French, one and all, put a single piece of clean writing paper between their hat and head during the hot months, to ward off the attacks of the *coup de Soleil*.

These are the diseases, which occur during the hot seasons; there is likewise a fever, in which the patient is continually affected with defluxions of the head: this appears in the late winter months, and during a wet spring; it is called a Catarrhal Fever: this disease is not frequent, but when it appears, it is generally treated like other fevers, except that bleeding is more freely used.

The pleurisy also makes its appearance sometimes in winter. Moderate or copious bleedings from the arm according to the degrees of violence of its attacks are immediately used: if looseness and gripes attend the pain, blood is taken away often, and in small quantities; the patient is kept moderately warm, and on no account suffered to uncover; the first medecine is commonly a cooling purge; gentle sudorifics are likewise administered; frequent hot baths for the feet are also prescribed, but very cautiously applied for fear of his catching cold. After the operation of purges and sudorifics, gentle antimonials are used, and a light easy digested diet, with infusions of hyssop, sage, or baum, follow in course; likewise swallowing of living wood-lice: and in case of costive symptoms, clysters are used; on a continuance of the pain in the side, a moderate blister or drawing plaister is put to the part; — much coughing, which causes a watchfulness, is removed by opiates: in feverish symptoms the disease is treated as the other fevers; spirituous liquors are to be avoided by all means.

During some winters, a *Peripneumony* also visits a few people here; the method of cure is the same as for the pleurisy: it is said to be more dangerous than the pleurisy, particularly if copious bleeding is not made use of as soon as the patient is affected. In this disorder there is generally a freer access of air allowed than in the last, and the patient kept almost in a sitting posture; it is said that the steams of warm water drawn into the lungs in this disease, is a powerful help.

A compound of the two last disorders, called the *Pleuroperipneumony*, is likewise sometimes heard of, and is treated as the last.

In Georgia I saw one or two instances of a disorder among blacks, to which the people give the odd name of the pleurisy of the temple, of the forehead, of the eye, and so on; I am told they have a pleurisy for every part of the head. It is violently acute, and, as I am informed, proves sometimes fatal in ten or twelve hours time; if immediately on its attack, a quantity of blood is not drawn from the arm, for the rest this disease is treated like a pleurisy.

The chronic diseases are dropsies, consumptions, hemorrhoidal and habitual fluxes, relaxed and bilious habits of body, ruptures, worm-fevers, and among blacks the leprosy, elephantiasis and body yaws; which last in Carolina is called the lame distemper; the first five of these are often best removed by a change of air, as the most efficacious medicines often prove of no use against the obstinacy of the disorders in the climate where they first originated.

The dropsy most frequently seizes a patient after an obstinate intermitting fever, where the use of the bark has been too long delayed: in this disorder the ordinary prescriptions in these countries is syrup of squills, and the common diuretic salt; with these the patient is confined to dry food, and from spirituous liquors; such vegetables as turnips, radishes, &c. he is allowed to indulge in: Dr. Lind says, that exciting a slight salivation, may be of help in a tolerable sound constitution, perhaps none of the chronic diseases are more relieved by change of climate than an obstinate dropsy. A consumptive habit of body, particularly where the cough is very obstinate and frequent, and when bilious stools, with a great hardness of the lower belly affect the patient, or when a continual fever emaciates the poor sufferer, he is in a dangerous way, and a remove to colder climates is hardly advisable; I have known such people relieved by making frequent short voyages to sea in moderate climates; but unless proper remedies are also made use of during these voyages, the fever returns almost directly on relanding; frequently after one or two of these voyages the patient feels himself better; if he then retires to a milk diet, and freely indulges himself in fruits, utterly avoiding all manner of drugs or medicines, he may find relief, and even a return of constitution; frequent doses of flour of brimstone, and cooling the water he drinks with sal nitre, are of use during this course: likewise the patient ought with the greatest care to avoid exercise; the stiller he keeps himself the more hopes of recovery there is. The fever which attends this disorder is of such a nature, that here the use of the bark must be carefully shunned, as it has been during long practice, and by frequent experiments of very able physicians, found to be a sure poison in this disorder. The Spaniards wear the nest of the great travelling spider sowed in a rag about their necks as a sure way to assuage a hectick fever, and I think with great success: it is a matter of surprize to see how perhaps a thousand *animalculae* which are in perfect life in one of these nests, at the time of its being put round the patient's neck, will in the course of about thirty hours be perfectly pulverized by meer dint of the heat of the body, which these young Spiders seem in a peculiar manner to attract. In hardness of the belly in this disorder most of the Creoles use hard and frequent rubbing it with a warm hand dipt in oil or hogs-lard, *Quere* bears oil being so very subtil and penetrating, would it not be preferable in rubbing?

The *haemorrhoidal* flux is very frequent here, and was it not so very troublesome an attendant, it would be looked upon as a beneficial event. Persons who are attacked by it are generally certain of not falling into the more dangerous diseases occasioned by obstructions of the *Viscera* in hot climates; the greatest danger attending it is that of the patient's falling into an habitual flux, which is a most tedious and troublesome disease, and that although the patient has no other complaint but the frequent necessity of going to stool and is but seldom troubled with an involuntary expulsion of the *faeces;* this disease is almost always a slow though sure harbinger of death by its continuance for years, draining to the very last drop of moisture from the sufferer, who being left a meer skeleton, is as it were carried off in the manner of an expiring candle-snuff; yet those persons, who have been opened after death have been found with all the inward parts perfectly sound, and thus the faculty is left in the dark without any way to account for this disorder; I have heard of people of a very robust constitution with whom it has continued above twenty years; no disease is so frequent; it almost always attacks people who have suffered much from frequent sickness or severe fatigue; and its obstinacy is such, that it will yeild to no remedy whatever in the climate where it originated; I have myself been attacked by it first in the province of Georgia, in consequence of the great fatigue I underwent in my frequent long and wearisome journies by land; no astringent of any kind, not even the long use of rhubarb and *ipecacuanha* was on any the least service to me; vain was every medicine against this obstinate malady; opium was recommended to me as a specific; this I took at length in incredible doses but the relief was only momentary; after the short reprieve obtained hereby it returned with tenfold violence and obstinacy: if then I was unhappy enough to use opium during this attack it was of no use whatever, but obliged me for the next time to seek respite from a double dose, the cold bath I found of some slight benefit and when I was at the proper season in any part of Florida, where the cocoplumb* grew in abundance, by freely eating this wholesome fruit I was relieved for that season, and no sooner was I obliged to abandon this excellent remedy, but the disease again prevailed. Thus was I harrassed for about eight years when I changed climates by coming to New-York; here likewise all medical prescriptions failed, till at length I found that a decoction of the bark of *Semi-Ruba* and *Terra japonica* in the proportion of half an ounce of each to six pounds of water being boiled down to one sixth, was an effectual medicine after the change of climate, which last alone must not be relied on: one quart of the above decoction in the quantity of a wine-glass full taken morning and evening cured me; but relapsing again after about three months I got another quart, with the two first glasses of which I took a small pill of crude opium, and by two more glasses full I found myself again restored to my natural habit of body.

An entire relaxation of the solids, and a bilious habit of body is another common affliction of those, who have suffered much by the diseases of hot climates; the constitution is in such people so decayed, that it seems as if every moment would be that of dissolution: the stomach is weak, their complexion is nearly that of a sufferer by the jaundice, and hardly any food especially greens and salads are found digestible; if the dry belly-ach has been their frequent attendant a paralytic contraction of the limbs is the final consequence of that malady: others again will frequently vomit clear bile and be very costive having the *abdomen* exceeding hard; for all these complaints there is no better cure than a change of climate, and when the patient begins to feel any benefit from the difference of air while at sea, I would recommend a plentiful and constant use of chamoemile flowers, chewing them in the same manner as people do tobacco; this, however disagreeable to most palates at first, becomes in time as agreeable to the mouth as it is grateful to the stomach. The use of

---

*Chrysobalanus.

Elixir Vitrioli in the quantity of fifteen or twenty drops taken every morning fasting and again an hour before dinner; and the moderate use of glass of generous wine is not amiss to such sufferers: animal food, especially mutton, is the most suitable diet, and in case of an obstinate costiveness use the *Elixir Aloes* often a night or in the morning; the cold bath, especially of salt water, is very beneficial to such sufferers.

Ruptures are pretty much complained of on the banks of Mississippi; I have observed likewise that they are a good deal frequent in Georgia and in Carolina: what can be the cause of a disorder of this kind being frequent I know not, but I find in a pamphlet which gives a superficial description of South-Carolina the following way to account for it, "the obstructed *viscera* being swelled beyond their natural size, the intestines are too much confined, and by nature of the ailment and bad digestion being frequently distended with wind, it is not to be wondered at, that they often pass through the rings of the abdominal muscles."

The Worm fever which is common through all America, especially from Pennsylvania southwards, is not so common here as in Carolina, Georgia, &c. the reason I take to be because the sweet potato is not so universally used for food here as elsewhere; children suffer most with it, though it sometimes affects people of all ages. When a fever obstinately withstands all medicines it may almost be depended upon, that this obstinacy proceeds from worms; the stincking weed which is known by the name of Jerusalem Oak, and in those provinces is the most efficacious vermifuge, and safest medicine especially for children; a spoon full of the expressed juice of the whole plant taken on any empty stomach is found to be a sovereign antidote; the *Lonicara* I have already mentioned as to its qualities; if the worms are suspected to be lodged in the *rectum,* clysters of a decoction of tansey, onions, garlick, rue, worm wood, and such like in milk are of good effect; a plaister of pulverized *Aloes,* oil of rue, or worm-wood, with powder of the bitter gourd and ox-gall applied to the navel is also of good effect: I would recommend the use of animal food, particularly rich fish-soups highly seasoned with garlick or onions, and it will be proper to avoid all kinds of farinaceous vegetables except wheaten bread: above all the potatoe and pumpkin ought to be shunned as poison.

A loathsome disease appears some times among the Negroes

after severe acute disorders, especially if the patient has been obliged to keep his bed long, likewise after a violent exercise has brought on a surfeit: this is called the *Elephantiasis* from the swelling of the feet and legs; it is most frequently seen to affect one leg only; in the first stages of this disorder the patient becomes wretched through excessive lassitudes which bring on an emaciation of the body, then the corrupted juices subside into the leg or legs and feet, these swell, the skin becoming distended, shines and shews the distended veins every where below the knee; now the skin by degrees loses its gloss and becomes unequal and something scaly; after this chaps make their appearance, the glands are stretched and the scales are daily enlarged, appearing as hard and callous as the hide of an Alligator, notwithstanding which the slightest prick of a pointed instrument will cause the blood to exude; this disease affects neither the appetite nor the digestive powers of the body, on the contrary the patient in this and chearfulness of spirits resembles the healthiest of men, and the inconvenience of his heavy leg only prevents his ability for the more laborious part of his duty.

No manner of cure has yet been found for this cruel disorder, but the patients often live to a very advanced age under the pressure of its yoke, even when it has been contracted in early youth; it is said that the amputation of the affected limbs is no cure, for the disease will immediately attack the sound leg; this I find also asserted by *Hughes* in his Natural History of Barbados.

I have seen three or four instances of the disease called body yaws (in the Islands) and in Carolina the lame distemper, this is said to proceed from hereditary venereal taints; it appears in cancerous corroding sores in the mouth and throat, and spreading ulcers together with fleshy protuberances chiefly on the face, breast and thighs, with a swelling of the skin and knee-bones, and commonly corrodes the Cartilages of the nose, its first symptoms shewing themselves about the throat and palate, have caused ignorant people to mistake it for the *Angina Suffocativa* before described: Mercurial medicines are used against it, afterwards diet-drinks of China root, nut-grass, &c. the sores in the mouth are often to be rubbed with a feather dipt in syrup of roses to an ounce of which two drops of *Sp. Vitr.* have been added; unctuous, salt, spiced meats and spirituous liquors are absolutely to be avoided; frequent sweats are also prescribed and a great care against catching cold.

The Leprosy so called, whether the same as was the cause of proscription to the unhappy patients under the Mosaic laws I shall not pretend to determine; certain it is, that it is a nauseous, loathsome and infectious disease some times seen among the blacks; this appears first with the loss of beard and hair from the eyebrows, swelling of the lobes of the ears, the face begins to shine and brown protuberances appear thereon, the lips and nose swell to a monstrous size, the fingers and toes will in the end drop off, and the body becomes at last so ulcerated as to make the poor incurable patient really a miserable object of pity.

Having thus largely described the climate, soil, water, general productions of the earth, the inhabitants with their customs, manners and the diseases incident to the human race here, I shall next proceed to give a topographical description of the country, the general face particularly, and with this subject end my first volume.

The river St. Mary is the northernmost boundary of East Florida and by the Spaniards called Rio *Santa Cecilia*, and by the savages *Thlathla Tlakuphka*, and is described in page 36; the soil here is not very fertile, unless it be at its very head, which reaches up to near the head of flint river, which last runs into the river *Apalachicola*, and this being the western bounds of this province, we may in the present undetermined state of these matters regard as the most natural limits of the province on the land side; from St. Mary's river's mouth there is a distance of fourteen miles along the beach of *Amelia* island to the mouth of Nassau river, this island is in general sandy and hilly, but has some fertile spots on it capable of fine improvements, it is about two miles through in its broadest parts, which is the north end, and tapers away till it's south end is scarcely half a mile across; here near the south end was formerly a convent of Nuns and a church dedicated to St. Mary, whose name the island bore in the time of the Spaniards, but there were scarce any traces left of the buildings in 1770; west of its north end lay the Tyger islands,between which and *Amelia* there is a tolerable broad passage forming a convenient harbour, these islands consist chiefly of marsh and pine hammocks; this passage continues through *Amelia* narrows into Nassau river, and is frequented by such crafts as exceed not four feet draught of water. The land on Nassau river is almost all very fertile; the river originates not above thirty miles from the sea, where it begins in many small rivulets, which all joining very soon, by their confluence form a considerable large stream navigable for vessels of ten feet draught up to its forks, this river formerly called Rio *Sta. Maria*, in my opinion has along its banks the best body of land in East Florida near the sea, and I believe that this will form in time a pretty opulent district; between this river and St. John's river the land is miserably barren, as it is likewise between Nassau and St. Mary, though not in so wretched a degree; we find between the mouths of Nassau and St. John's river three islands, great and little Talbot, and fort George, together in length about five or six miles; the Talbot's islands lay next to the south of *Amelia*, and the greatest innermost: this last is counted fertile, the other is inconsiderable; they are hilly as well as fort George; the last has received its name from an intrenchment thrown up on it by general Oglethorp during his fruitless expedition against St. Augustine.

We are now at the mouth of the river St. John, already in part described in page 34 and 35; this river called by the indians *Ylacco*, and by the Spaniards Rio *St. Matheo o Picolato*, runs from its mouth something above twenty-four miles up nearly east and west; at this distance up the ferry is kept: (commonly called the cow-ford from the multitude of cattle drove through here at the beginning of the settlement by the English:) just above this ferry the river takes a bend and the direction of its course becomes parallel to the ocean; the general run of its water being from south to north rising at or near the latitude 27: and this ferry laying nearly in 30: 10; near the latitude 28: it approaches the sea within a mile, in so much, that some marsh and broken pine-land only separates them: thus it makes a peninsula from hence to the above latitude 30: 10. which peninsula is for the most part between twenty-five and twenty miles in breadth. This part of the country is the chief seat of the present improvements, and is for the most part too barren for planting; but excellent for the keeping of cattle and horses, as the land chiefly consists of pine-land and savannahs of the kind described in page 22: the swamps along the river (although their appearance from the river deceived the first adventurers,) are for the most part no more than an edging along its banks, hardly any where extending a quarter of a mile deep; the journal of Mr. *Bartram* as published by Dr. *Stork*, may give a tolerable idea of the banks of this river, and consequently of the west part of this peninsula; but this journal, though a very loose performance, and principally defective where we might expect it most compleat, viz. in the botanical articles, yet such as it is, the Doctor who wanted to extol this province even beyond reason, has not thought fit to give it us in its native dress, but mutilated and unfairly modelled it to answer his own purpose, which has given another author* a handle for depreciating this country still more below its value than the Doctor has endeavoured to raise it above; however all such prejudiced writers being below contempt, let us leave them what they are, and after observing once more that the fruitful spots are few when compared with the unfruitful in this tract, proceed to give an account of the eastern edge of this peninsula viz. the sea coast from the mouth of St. John to the mouth of the river *Ais* or *Aisahatcha:* not far from the entrance of St. John's in *St. Pablo* or St. Paul's creek or river; on it are some fruitful spots about twelve miles south of the great river; *Sablo* river approaches the head of St. Mark's or the North river, at whose head are the plains of *Diego* very proper for the breeding of cattle. This river St. Mark falls into the harbour of Augustine about two miles to the northward of the fort, and but a little way within the bar: to which bar from the general's mount at the entrance of St. John's is a measured distance of 36 miles. Between the north river and the sea is a very narrow slip of sand-hills and other almost useless ground; the bar of this harbour is a perpetual obstruction to St. Augustine's becoming a place of any great trade, and alone is security enough against enemies; so that I see but little occasion for so much fortification as the Spaniards had here, especially as a little look out called *Mossa* at a small distance north of the town, proved sufficient to repeal general Oglethorpe with the most formidable armament ever intended against Augustine; however there was much more propriety in the Spaniards having a fort in the modern taste of military architecture of a regular quadrangular form, with four bastions, a wide ditch, a cover'd way, a glacis, a ravelin to defend the gate, places of arms and bomb-proofs with a casemating all round &c. &c. for a defense against the savages, than there was in raising such stupendous piles of building as the new barracks by the English, which are large enough to contain five regiments, when it is a matter of great doubt whether there will ever be a necessity to keep one whole regiment here; to mend this matter the great barrack was built with materials brought to Augustine from New York, far inferior in value to those found on the spot; yet

---

*Present State of Great Britain and North America.

the freight alone amounted to more than their value when landed; so that people can hardly help thinking that the contrivers of all this having a sum of money to throw away, found a necessity to fill some parasite pockets, and judged the best method for doing this was to make contractors of the folks in view, as fifty men besides the inhabitants would defend this fort against the united bands of all the savages here; as a constant supply of provisions can come by water, and no formidable foreign force can ever be reasonably expected to come here, it makes us almost believe that all this shew is in vain, or at most, that the English were so much in dread of musketos, that they thought a large army requisite to drive off these formidable foes. To be serious, this fort and barracks add not a little to the beauty of the prospect, but most good men will think with me that the money spent on this useless parade would have been better laid out on roads and ferries through the province, or if it must be in forts, why not at Pensacola? where there is a necessity to have a powerful force, and where the bar will admit a sixty gun-ship: but I had almost forgot that the acts of Assembly there would have defeated the sanguine expectations of Messrs. contractors.

The town has by all writers till Dr. Stork's time been said to lay at the foot of a hill; so far from truth is this, that it is almost surrounded by water, and the remains of the line drawn from the harbour to St. Sebastian's creek, a quarter of a mile to the north of the fort, in which line stands a fortified gate called the barrier gate, is the only rising ground near it; this line had a ditch, and its fortification was pretty regular; about a mile and a half beyond this are the remains of another fortified line, which had a kind of look out or advanced guard of stoccadoes at its western extremity on St. Sebastian's creek; and fort Mossa at its east end; besides these the town has been fortified with a slight but regular line of circumvallation and a ditch. The town is half a mile in length and its southern line had two bastions of stone, one of which (if not both) are broken down, and the materials used for the building of the foundation of the barracks; the ditch and parapet are planted with a species of agave, which by its points is well fitted to keep cattle out, and with that intention was used by the Spaniards for fences. Dr. Stork has raised this fence into a fortification against the savages and magnified it into *Chevaux de frize*. The town is very ill built, the streets being all except one crooked and narrow. The date on one of the houses I remember to be 1571; these are of stone, mostly flat roofed, heavy and look badly. Till the arrival of the English neither glass windows nor chimneys were known here, the lower windows had all a projecting frame of wooden rails before them. On the 3d of January 1766, a frost destroyed all the tropical productions in the country except the oranges; the Spaniards called this a judgment on the place, for being become the property of Hereticks, as they never had experienced the like. The Governor's house is a heavy unsightly pile, but well contrived for the climate; at its north west side it has a kind of tower to the height of which Governor Grant has added several feet; this serves as a look out. There were three suburbs in the time of the Spaniards, but all destroyed before my acquaintance with the place, except the church of the Indian town to the north now converted into an hospital; (Dr. Stork says the steeple of this church is of good workmanship though built by the indians, neither of which assertions is true,) and the steeple of the German chapel to the west of the town likewise remains. The parish church in the town is a wretched building and now almost a heap of ruins; the parade before the Governor's house is nearly in the middle of the town and has a very fine effect; there are two rows of orange trees planted by order of Governor Grant, which make a fine walk on each side of it: the sandy streets are hardened by lime and oyster-shells; Dr. Stork says there were nine hundred houses at the time of the Spanish evacuation, and three thousand two hundred inhabitants; in my time there were not three hundred houses and at most a thousand inhabitants; these, a few excepted, I found to be a kind of outcast and scum of the earth; to keep them such, their ill form of government does not a little contribute. A letter from a friend, who lately returned from England, directed to me, dated 2 th May 1774, says, "this town is now truly become a heap of ruins, a fit receptacle for the wretches of inhabitants."

About eight miles north of the town is a tract of a considerable size, called the twelve mile swamp; which I take to be equal in goodness to any in America; this began to be cultivated in the year 1770. Three miles west from the town is another small tract of tolerable land, called the three mile swamp; all besides this within many miles is really a miserable sand, a dreary scrub ground, and boggy salt marsh, a few acres at the head of St. Sebastian's only excepted.

The island *St. Anastatio* is situate opposite the town and forms the harbour, its north end being opposite to the south end of the narrow peninsula made by the north river: this is the only part of all the lands purchased by the worthy Mr. Fish, for the debts due to him by the Spaniards, which he was allowed to keep, notwithstanding the treaty allowed it him; for no sale was judged valid if made by him, unless the government afterwards gave a grant, but this was one of the many ways to make offices worthy of holding; the above named gentleman has a pretty retreat on this island, about four miles from town; but as the land is barren it is more pleasant than profitable. The sound which divides this island from the main, is called *Matauca* river, it is about twenty miles in length; the island affords pasturage for numerous herds of horses and some cattle, the first in particular breed kindly here; about half a mile from the north end of the island is a heavy stone building serving for a look out; a small detachment of troops is kept here, and by signals from hence, the inhabitants are given to understand what kind of, and how many vessels are approaching the harbour, either from the north or from the south; in the year 1770 fifty feet of timber frame work was added to its former height, as was likewise a mast or flagstaff forty seven feet long, but this last proving too weighty, endangered the building and was soon taken down. In this island are quarries of a kind of concreted petrified shells, or a stone bearing their resemblance, this will be described hereafter.

About two miles south of the town is the mouth of St. Sebastian's river already mentioned, this is a small creek rising near six miles north of the town, but its stream soon becomes salt by the meeting of the waters from below; vessels go two miles and a half up it to clean. A bridge was built over it to the west of the town, which shortened the road into the country near seven miles, but the great depth of the water joined to the instability of the bottom did not suffer it to remain long, and a ferry is now established in its room; the keeper of the ferry has £ 50 sterl. *per annum* allowed him, and the inhabitants pay nothing for crossing except after dark.

We next meet the mouth of St. Nicholas creek, on the point to the north of which the first town was built by the Spaniards, but they soon removed it for conveniency's sake to its present site: on the south point is a plantation belonging to Mr. *Moultrie* the present Lieutenant Governor; this place is remarkable for a large oyster bank being entirely eat out by this gentleman's negroes in a scarce season. This river is short originating about ten miles to the west of the sound, and has no fertile land on or near it.

Six miles further south, or about ten miles from town is the mouth of *Sta. Cecilia*, another small river of little note. Between nine and ten miles further is the look out or fort of *Matanca*, on a marshy island, commanding the entrance of *Matanca*, which lays opposite to it; this fort is to be seen at the distance of about five leagues, it is of very little strength, nor need it be otherwise, as there is scarce eight feet water at the best of times on this bar: the Spaniards kept a Lieutenant's command here; the English a Serjeant's. Between two or three miles from this inlet or bar is another of still less note, called *el Penon*; opposite and southward is the mouth of a river called *North West*, I suppose from the direction of the course we go up it; it is somewhat more considerable than the two or three last mentioned; some settlements are near its banks, but of small importance, the land being in general arid and poor. Three miles further south is the upper end of the water which by former Geographers has been stiled a river; the Spaniards called it *Rio Musketo*, and the English the Musketo river: *Mr. de Brahm* has changed the name of it into *Halifax* river. This pretended river is one of those arms of the sea commonly called a *Lagoon;* thus I believe the wonder of Dr. Stork's two rivers, lying parrallel to each other and the ocean, and yet running directly opposite courses is well accounted for, such *Lagoons* being not at all uncommon; the upper part of this piece of water is a kind of lake or pond, and is removed at so inconsiderable a distance from the sea, that boats which were obliged to go outside at *Penon,* where the inland navigation ceases, are here drawn out of the sea into this *lagoon.* I believe the first rocks to be seen on the beach of the coast of America, from Long-Island to this place are to be seen here; they are of that kind which seems to be a concretion of shells; they are small and do not extend far into the water, but are as it were buried in the sand at low water mark.

About twenty miles south is the mouth of *Tomoco* creek, falling into this *lagoon* from the westward; the head of this is at a small distance from St. John's river; it is said that some good land lays along its banks.

Thirty two miles further south is the mouth of another river called *spruce creek;* it is very rich in fresh marsh, the ground being greatly broken at its mouth; this is a more considerable stream than the last, and at a little distance from this mouth are the first settlements of *Musketo or New Smyrna*, particularly *Mr. Penman's*.

Between three and four miles further south is the entrance, or bar of *New Smyrna* of still less navigable convenience than St. Augustine, and is the outwatering of the above named *lagoon* as likewise of another coming from the south stiled *North Hillsborough river*; here we begin to see a few of the tropical plants, such as *carica, borassus, capsicum, mangles* and *blackwood*. At a few miles from the bar is the situation of the town or settlement made by Dr. Turnbull for Sir William Duncan, himself, and perhaps more associates; this town is called *New Smyrna*, from the place of the Doctor's lady's nativity. The settlements round this famous town extend considerably along the banks of this *lagoon*, and large quantities of very good indigo have been made here. If my reader is inquisitive to know why I call this *famous*, I answer on account of the cruel methods used in settling it, which made it the daily topic of conversation for a long time in this and the neighbouring provinces.

About 1500 people, men, women and children were deluded away from their native country, where they lived at home in the plentiful cornfields and vineyards of Greece and Italy, to this place, where instead of plenty they found want in its last degree, instead of promised fields, a dready wilderness; instead of a grateful fertile soil, a barren arid sand; and in addition to their misery, were obliged to indent themselves, their wives, and children for many years, to a man who had the most sanguine expectations of transplanting *Bashawship* from the Levant. The better to effect his purpose, he granted them a pitiful portion of land for ten years, upon the plan of the feodal system: this being improved and just rendered fit for cultivation, at the end of that term it again reverts to the original grantor, and the grantee, may, if he chuses, begin a new state of vassalage for ten years more. Many were denied even such grants as these, and were obliged to work in the manner of negroes, a task in the field; their provisions were at the best of times only a quart of maize per day, and two ounces of pork per week; this might have sufficed with the help of fish which abounds in this *lagoon*, but they were denied the liberty of fishing, and lest they should not labour enough, inhuman task-masters were set over them, and instead of allowing each family to do with their homely fare as they pleased, they were forced to join all together in one mess, and at the beat of a vile drum, to come to one common copper, from whence their *homany* was laded out to them; even this coarse and scanty meal was through careless management rendered still more coarse, and through the knavery of a proveditor, and the pilfering of a hungry cook, still more scant. Masters of vessels were forewarned from giving any of them a piece of bread or meat. Imagine to yourself an African (an expert hunter) who had been long the favorite of his master, through the importunities of this petty tyrant sold to him,—imagine to yourself one of a class of men, whose hearts are generally callous against the softer feelings, melted with the wants of some of these wretches, giving them a piece of his venison, of which he caught what he pleased, and for this charitable act disgraced, whipped, and in course of time used so severely that the unusual servitude soon released him to a happier state; again, behold a man obliged to whip his own wife in public, for pilfering bread to relieve her helpless family; then think of a time when the above small allowance was reduced to half, and see some brave generous seamen charitably sharing their own allowance with some of these wretches, the merciful tars suffering abuse for their generosity, and the miserable objects of their ill-timed pity, undergoing bodily punishment, for satisfying the cravings of a long disappointed appetite, and you may form some judgment of the manner in which New Smyrna was settled. Mr. Joseph Purcell, an excellent young man, who was draughtsman to our department, a Minorquin, who with his family came over at the same time with these people, but happily withdrew from the yoke, could never speak of this without tears; he had been several times an eye witness to this distres, and told me, that he knew many among the unhappy sufferers who were comfortably established in Europe, but by great promises deluded away; and O Florida! were this the only instance of similar barbarity which thou hast seen, we might draw a veil over these scenes of horror; but Rolles Town, Mount Royal, and three or four others of less note have seen too many wretches fall victims to hunger and ill usage, and that at a period of life when health and strength generally maintain the human frame in its greatest vigour, and seem to insure longevity. Rolles-Town in particular has been the sepulchre of above four hundred such victims. Before I leave this subject, I will relate the insurrection to which those unhappy people at *New Smyrna* were obliged to have recourse, and which the great ones stiled rebellion. In the year 1769, at a time when the unparalleled severities of their task-masters, particularly one *Cutter* (who had been made a a justice of the peace, with no other view than to enable him to execute his barbarities in a larger extent, and with the greater appearance of authority) had drove these wretches to despair, they resolved to escape to the *Havannah*; to execute this, they broke into the provision stores, and seized on some craft lying in the harbour, but were prevented from taking others by the care of the masters. Destitute of any man fit for the important post of a leader, their proceedings were all confusion, and an Italian of very bad principles, who was accused of a rape on a very young girl, but of so much note, that he had formerly been admitted to the overseer's table, assumed a kind of command; they thought themselves secure where they were, and this occasioned a delay, 'till a detachment of the ninth regiment had time to arrive, to whom they submitted, except one boat full, which escaped to the Florida Keys; but was taken up by a Providence-man: many were the victims destined to punishment; as I was one of the grand jury which sat fifteen days on this business, I had an opportunity of canvassing it well, but the accusations were of so small account that we found only five bills; one of these was against a man for maiming the above said Cutter, whom, it seems, they had pitched upon as the principal object of their resentment, and curtailed his ear, and two of his fingers;—another for shooting a cow, which being a capital crime in England, the law making it such was here extended to this Province; the others were against the leader, and three more, for the burglary committed on the provision store; the distresses of the sufferers touched us so, that we almost unanimously wished for some happy circumstances that might justify our rejecting all the bills, except that against the chief, who was a villain. One man was brought before us three or four times, and at last was joined in one accusation with the person who maimed Cutter; yet no evidence of weight appearing against him, I had an opportunity to remark by the appearance of some faces in court, that he had been marked, and that the grand jury disappointed the expectations of more than one great man. Governor Grant pardoned two, and a third who was obliged to be the executioner of the remaining two. On this occasion I saw one of the most moving scenes I ever experienced; long and obstinate was the struggle of this man's mind, who repeatedly called out, that he chose to die rather than be the executioner of his friends in distress: this not a little perplexed Mr. *Woolridge,* the sheriff, till at length the entreaties of the victims themselves, put an end to the conflict in his breast, by encouraging him to the act. Now we beheld a man thus compelled to mount the ladder, take leave of his friends in the most moving manner, kissing them the moment before he committed them to an ignominious death. I have dwelt the longer on this subject, because the native prejudice of vulgar Englishmen, has represented the misfortunes of these wretches in too black a light. It is said that Dr. Stork, who was near the spot when the insurrection happened, died with the fright, and Cutter some time after died a lingering death, having experienced, besides his wounds, the terrors of a coward in power, overtaken by vengeance.

To return to our *topographical* account:—along this southern lagoon, which extends itself for about forty miles to *Cape Cannaveral,* in the latitude of 28 degrees and a half, we find several settlements of good note, among which that of Captain Rogers is the most meridional habitation on the British continent. At *Cape Cannaveral* is some good plantable land, and here is the southern head of this lagoon; about two miles and an half to the westward thereof is the head or northern end of another branch, likewise called a river; a road is cut to draw boats out of the Musketo Lagoon into this, which is called *South-hillsborough* by *De Braham,* but commonly called *Indian River;* the savages call it *Aïsa Hatcha,* i.e., Deer River, although the same elegant *Hexiphanes* has made it *Hysweeslake;* a word by him fabricated, as I suppose, from *Ylacco,* the name given by the savages to St. John's River; the Spaniards call it *Reo d'aïs.* No rivers of any note fall into its northern branch, except St. Sebastians, directly opposite to whose mouth happened the shipwreck of the Spanish Admiral, who was the northermost wreck of fourteen galleons, and a hired Dutch ship, all laden with specie and plate; which by stress of north east winds were drove ashore and lost on this coast, between this place and the bleach-yard, in 1715. A hired Frenchman, fortunately escaped, by having steered half a point more east than the others. The people employed in the course of our survey, while walking the strand, after strong eastern gales, have repeatedly found pistareens and double pistareens, which kinds of money probably yet remaining in the wrecks, are sometimes washed up by the surf in hard winds. This *Lagoon* stretches parallel to the sea, until the latitude 27:20, wheer it has an out-watering, or mouth: directly before this mouth, in three fathom water, lie the remains of the Dutch wreck. The banks of this lagoon are not fruitful.

Having now exceeded the latitude 28, to the southward, I shall here break off from the seacoast, and resume the description of the interior country from that latitude; being, as before observed, the south-end of the peninsula, made by St. John's River and the sea. This southern end is a mere point of marsh, with some broken pine land in it, not much above three quarters of a mile wide, dividing the fresh water of St. John, from the salt of *Aïsa hatcha*: imagine then to yourself a country gradually rising into a ridge of highland, very barren, sandy and gravelly, a few places excepted, intersected with abundance of rivulets, and variegated with ponds and lakes, whose banks being in general lined with oak *Magnolia*, and other trees, exhibit the most romantic scene imaginable, and you will have a just idea of this place. We frequently meet with spacious savannahs of the high kind; the country is covered with roe deer and turkies, the lakes stocked with fish, and thus it continues in a due west line across the Mexican gulph to the said latitude 28, which strikes said gulph 15 m. northward of the bay of *Spiritu Santo*. As we go northerly the fertile spots become more frequent, till we come to that part of the province formerly possessed by the Spaniards: here we find along the road from *Augustine* to *Apalachia*, the remains, or ruins of the following forts and towns, viz. on each side of *St. John's*; fort *Picolata*, *Popa*, and about seventy miles W.N.W. of *Popa*, we meet with *Alachua*, *Puebla nova Navala*, *Santa Fe*, in a very fertile region, *Utoca*, *St. Pedro*, and *St. Matheo* in a less favourable soil; then some small remains of *Ayovola*; next the fort *St. Mark's* in pine barren, all which are still marked (even in late maps) though not now inhabited; in this part of the country, especially at *Alachua*, an hundred miles N.W. from *Augustine*; the ground is fertile in a high degree, and so continues northward, to the extent of the province. Near *Santa Fé* is a river of the same name, which by the *English*, is corruptly called *St. Taffy's*; this river, after a considerable course, loses itself under ground, and the geography is not sufficiently known to acquaint us whether it emerges again or not. This I reckon one of the chief curiosities of this province. We next met with very great holes, fifty or sixty feet deep, and thirty or forty feet in diameter at the surface, in the highest part of the pine ridges, which afford some good water to the thirsty traveller, there being no other water in those ridges. I came across the peninsula in a dry season, and we were five days without water, because we travelled over the ridge almost in its longitudinal direction, and saw but few of these holes; (in the southern latitude of 28 and 29, in which I was then) even these we found dry; but I did not know, at that time, their nature, which I afterwards found. My guide informed me, that in the road from *Augustine* to *Apalachia*, in a similar season, he had been obliged to suffer a long thirst, and at last came to one of these holes, in which he had always found water, though in others it failed; but this time he found his expectations frustrated; pressed by drought, and animated by his companions, they, with some labour, removed a flat stone, which lay at the bottom, with an intent to dig for water; but how were they surprised, when they saw a small hole, and about six inches below its surface some very fine water; having satisfied the call of nature, their contemplation led them to consider whence this water sprung; one of them imagined he saw it gently running, their staves were put in to feel for bottom, but none was found: inquisitive curiosity led them to sound deeper, and lo! a sapplin, of near thirty feet, found no bottom, and they departed with unsettled conjectures.

On the 10th of June, 1771, I found myself in a disagreeable situation, on the west coast of East-Florida, through want of water; I saw frequently very strong currents of discoloured water, indicating fresh streams; supposing these to proceed from rivers, I coasted along shore; but found nothing except innumerable islands and vast shoals; frequently tasting the water, found it always brackish: at length, on the 21st day of said month, I found an opening, where I got anchorage for my vessel. I went personally on shore, to explore the small island we were next to, taking with me a lad, and a thirty gallon cask. Its aspect was miserable; oyster shells and rocks were every where its superficies; at last, however, I found a small hole in the rock, and water in it; which, to my inexpressible joy proved fresh, and of the clearest and finest kind; hastening back on board, we were obliged to roll the cask in the water, in order to get it into the boat, when, accidentally applying my hand to my mouth, I found to my surprize the water to be fresh; having proceeded on board, we tasted the water in which the vessel swam, which also proved fresh and good: thus I found myself, as it were, in a sea of fresh water; stimulated by curiosity, I left the vessel next day, with an intention to penetrate to the main land, which, with difficulty was effected;

but in a whole day's search no river was found; yet a constant strong current proceeded from the shore, and along it. The singularity of the circumstance made me beat the ground, and stamp on it, which occasioned a hollow sound. I have since been informed by travellers, that between the town St. Mattheo, (now inhabited by savages, and situate on the river St. Juan de Guacaro) and Apalachia, they frequently meet with this hollow sound under their horses feet. Twenty miles east from Apalachia, we cross a place over a creek, called by the savages, the natural bridge: all these circumstances correspond with an idea which I have formed of subterraneous rivers in this province. Another curiosity, though not a natural one, are the marks of former improvement of this country; particularly the vestiges of the regular maize hills (even in woods where, since the aera of culture, trees of twelve to eighteen inches diameter have grown up) and the nails and spikes drove into some very large trees, apparently at ancient Spanish Cowpens; add to these, the reliques of old fences, huts, houses and churches, particularly a church bell in the fields, at Santa Fé. Thus we have a melancholy instance, even in this new world, of the depredations of time, and a country once nobly and extensively settled, through the inroads of the savages reduced again to a wilderness, yet all this must have happened some time later than Anno Domini 1543, the year when Soto compleated the first Christian expedition to Florida, with his death.

Tanta est terrestrium vicissitudo!

The remaining natural curiosities are the mineral springs, described in page (34) and (35), and the ridge of sand hills between *Oclawwawhaw*, and *St. John's*, likewise mentioned in page (35) and (36) of this volume. I also regard as a curiosity of the artificial kind, the immense orange groves, found in the woods between latitude 28½, and 30°, supposing them originaly sprung from the seeds of some oranges formerly dropt by travelling Spaniards, at their camps; but which are now gradually decaying, by reason of the wantonness of our traders and hunters, who, when in want of the fruit, cut down the tree; other vegetable curiosities will more probably appear under the botanical heads. The many *Tumuli*, originally intended for the peaceful repository of the dead, tho' some of their tops now bear summer houses for recreation to the living, are another curiosity. Having mentioned hills, I could wish to know whence comes the idea of *Apalachian Mountains*, nothing above a moderate hill appearing in all this vast extent of country, till we meet (to the north of latitude 35) with the mountains of the *Upper Cherokees*, on the *Tanassee* river; and it is observable, that this ridge continuing back of North-Carolina, Virginia and Maryland, through Pennsylvania, and back of New-Jersey, into New-York, and terminating near Katskill, about eight miles from Hudson's River (where they bear the name of Blue Mountains) lie nearer to the ocean than to the Mississippi; all the country west of them is nearly a plain, there being no falls from *Monongahela* and *Alligheny* forks, down to the mouth of Mississippi, that are of consequence enough to interrupt navigation; even the *Ohio* falls, so called, are not violently rapid; nor are there any falls from the place of embarkation on the *Tanassee*, down the *Hogoheegee*, to the *Ohio;* the *Shawanese* river, has none that ever I could learn; *Tombechbe* none in all its extent;—whence then comes the descriptions of *Appalachian Mountains?* whence *Augustine*, at the foot of a hill, and many more such dreams not worth repeating. Either the face of the country is more changed than any we read of elsewhere, or former writers having been abused, have again imposed on us; for the face of Greece and other regions of the East, is yet reconcileable to *Homer's* geography. *Holland* still has nearly the same kind of superficial appearance, as we are informed, it had eighteen hundred years ago; and we may still trace *Julius Caesar* through these ancient habitations of the *Catti, Kelti* and *Cherusti;* how then comes America so altered? Most of the describers of this new world, even late ones must have imposed upon us, either originally, or by copying such lying exaggerators and impostors, as the inventors of the conquest of *Mexico*, and others of his stamp; who, not content with giving us plain truth, and simple facts in their native dress, have almost hid them from us by the thick clouds of fable.

In this northern part of the province we find the following settlements of savages; about the middle of the land, nearly in latitude 28, in a village called *New Yufala*, being a colony from *Yufala*, in the upper Creek nation, planted in 1767, in a beautiful and fertile plain: about fifty miles N.E. from whence, and about thirty W.S.W. from the upper part of Lake George, in St. John's River, which is eighty miles nearly south of *Augustine*, is another town, situate on a very beautiful lake, about seven miles in length, and from one to one and a half in breadth; this town is accessible only on the N.W. side of the above lake, over a very narrow and high ridge: when I came to it from the southward, though in a very dry season, we had near a mile and a half, or two miles, to wade breast high, and sometimes deeper, through a kind of marsh on the west bank. The town lies near the N.W. end of the lake, and from its situation is called *Taloffa Ockhasé*, i.e. the town of the lake; some islands in this lake, as well as some land near the town, are cultivated by the savages. Further north is *Alacua, Santa Fé*, and *St. Mattheo*, now inhabited by the savages; the fort at *Apalache* is new a trading house, chiefly in their possession, and the western extremity of the province is the great river *Apalachicola*, on which the lower Creek nation is settled, whose principal towns are called the *Cowettas, Chatahoochas, Euchas, Citasees, Hogoleeges, Oakmulgos, Tuskeegies, Cussitos*, and *Cherokee Lousitsa*, near the ruins of fort *Apalachicola*, at the junction of Flint River, and the river *Apalachicola* in the south western extreme of this division is the head of *Manatee* river, between which and the *Amaxura*, I saw a vast number of deer, and the marks of many of the hunting camps of the savages. We found the foot steps of six or eight buffaloes hereabouts, so plain as to be convinced of the track being made by those animals, but saw none of the animals themselves; the *Amaxura* or some of its branches is not far from the *Manatee*, and where we crossed it, was an extensive piece of excellent land.

To compleat the *Topography* of this part, I shall mention that *Rio d'aïs* abounds so much in fish of various kinds, that a person may sit on the bank, and stick the fish with a knife, or sharp stick, as they swim by; and I have frequently shot from four to twelve mullets at one shot, nay our boys used to go along side of the vessel in the boat, and kill the cat-fish with a hatchet, or stick. This might well be ranked among the curiosities; but the whole country abounds so much in fish, that when *St. Augustine* flourished most, the fishermen frequently used to allow people, who brought a *Real*, to take as much as they pleased, out of their boats.

The reader will observe that the rules of geography, with regard to natural limits, have caused me to vary a little in the present division of the peninsula, from that adopted in page (1) and (2) of this work, which was founded in the climates; but the variation being small, I hope it is excuseable. I shall now proceed with the description or topography of the southern part of the peninsula.

From the latitude 28, the lagoon of *Aisa*, or Indian River, has not any thing very remarkable, until latitude 27: 20, where there is a mouth, or outwatering, into the ocean, with several small inlets within it. This mouth can seldom be entered by any vessel that draws above five feet water; and before it, in the sea, are two bars, the inner one having about ten, the outer one seventeen feet on it; and this outer one is near four miles from the land. The sand before this entrance, is a fine white quicksand, of a particular nature. I have anchored several times within the distance of three or four leagues from this mouth, but not above once or twice without having the cable all eaten through, in the ring of the anchor; sometimes I have purchased the anchor by a single strand, sometimes by less, at other times by a little more; and I have lost above six or seven anchors, and large grapnels at this place, yet there is nowhere any foul ground, or, in other words, no rocky bottom. At a mile, or more, from the shore, and within that distance, I think, I have found a few shell stones, I imagine that this fine quicksand, being very sharp, by its continual motion chafes and frets the cable quite through, and this, it generally does, in less than four and twenty hours.

From this mouth of the lagoon an island stretches to about the latitude 26, 55, where there is another mouth, or inlet, caled *Hobé* by the Spaniards; this island is thirty-nine measured statute miles long: at twenty-four miles from its north-end are several high cliffs of a blue stone; these are the first rocks that lie high out of the water, along the American beach; they are placed at about high water mark, and a small ridge, or reef, runs sloping off from the northernmost one; and about nine miles further, towards *Hobé* is another parcel; as there is likewise at the entrance of *Hobé*, of very solid, hard rock; all of which, particularly the northernmost ones, are excellent landmarks for the seamen, going southwards. The remarkable things about this island are, that it is a narrow slip of beach, and mangrove land; on the beach are always great numbers of pieces of Spanish cedar; originally cut on the windward rivers of Cuba, for the use of the *Catholic* King's ship-yards; but by land floods drove into the Bahama channel, and Gulph stream, whence the frequent east-winds force them upon soundings, and so on this beach; very few pieces are found either north or south of this island. The island on its westside is indented, almost regularly into points and bays. Fresh water may be obtained by digging in almost every part the beach. The chief growth here are *mangroves, blackwood, conocarpus, salsola* and *uniola;* with a parasitical plant of the genus *Tillandsia,* called wild pine apple. A few spots of *hammock,* or upland, are found on this island; these produce the *zantoxylum, sicus citri folio, Coccoloba, Mastick, Borassus,* and a few trees of the live oak, and willow oak, the *Chrysobolanus,* and the *Cercus Triangularis,* with that kind of *Cactus,* commonly called *Opuntia.* During the season the logger-head turtles land here in vast multitudes, to lay their eggs, which the bears profit by; for, led by instinct, or otherwise, these animals come in droves, and dig the eggs out: at this business they are so expert, that they dig wells for their supply of water; during their stay, they sometimes fetch down the wild *Pine,** which, by its structure naturally contains a considerable quantity of rain-water, preserved in a fresh, sweet taste. So industrious are the bears as digging up the eggs, that the turtle seldom leaves her nest above a quarter of an hour before they are eaten, insomuch, that a traveller, if he choses any of this provision, is obliged to watch the turtles coming. I have seen the bears approach within five or six yards of our camp, at times when we had some of these eggs, but this stretch of *latronical* boldness, generally cost them their lives.

Opposite this island the shore is sandy and high, here are many rivulets, and the *Berassus* here appears to be the master of the soil, scarce any other plant shewing itself; directly opposite the rocks before described is the mouth of *St. Lucia* river, which has a wide kind of bay, for eight miles up, stretching first N. E. 40 W. four miles, then W. four miles more, when it divides into two branches, one coming from the south, the other from the N. W. by N. this last appearing the principal, I went up it for twenty-four miles, reckoning direct distance, and found no where less than seven feet water; here the river became narrow, and partly on account of the obstructions by logs, partly on account of the rapidity of the stream, I left the vessel; and going up by land, found the river at last to run through a vast plain, the bank of the stream only being fringed with a few trees.

Here we shot what number of deer, and turkies we pleased, and might have continued so to do, I dare say, two months longer; the reason of this plenty is, in my opinion, that this tract is scarcely ever invaded by the hostile savages, or yet more destructive white hunter; this is the river, which, as I was told by a Spanish pilot and

*Tillandsia Lingulata.

fisherman of good credit, proceeds from the lake *Mayacco*, a lake of seventy-five miles in circumference by his account. The man told me that he had formerly been taken by the savages, and by them carried a prisoner, in a canoe, by way of this river, to their settlements on the banks of the lake; he says, that at the disemboguing of the river, out of the lake, lies a small cedar island; he also told me that he saw the mouth of five or six rivers, but whether falling out of, or into the lake, I could not learn of him; probably some of the many rivers I crossed in my journey across this peninsula, fall into it, and it is not improbable that St. John's river originates in it. The large river in Charlotte harbour, by the direction of its course, meridian situation, and great width, I judge, might, perhaps, spring from the same fountain; however, the savages of *Taloff Ochasé* told me, that in going far south, they go round a large water, emptying itself into the west sea, i.e. gulph of *Mexico*.

Thus much have I been able to learn of this water, the exploring of which I always intended; whether there is really this lake, or not, I will not be positive, but the above circumstances, joined to a dark account, which the savages give of going up St. John's, and coming down another river, to go into some far southern region of East Florida (on which account the name of *Ylacco*, and the name given to *St. Lucia* by the savages, both conveying indecent meanings, are by them given to these rivers) seems to confirm it. That there is some such great water, is further to be gathered from the profusion of fresh water which this river, *St. Lucia*, pours down. Such is the immense quantity that the whole sound between the abovenamed island and the main, though an arm of the sea, situate in a very salt region, and in general two miles wide, is very often rendered totally fresh thereby: in so much, that it has made the very speculative Mr. *De Brahm* insist upon having seen mangrove stumps in fresh water. This lake has given rise to the intersected and mangled condition in which we see the peninsula exhibited in old maps. But to return to the mouth of the river *St. Lucia*, it lays one mile, seventy four chains, and seventy links S.W. by S. from the great rocks; the mouth is fifty four chains, eighty-nine links wide; six miles and a quarter N. N. W. from the mouth on the edge of the sound, lieth the hill by the Spaniards called *Ropa Tendida*, and by us the *Bleach-Yard*, on account of its appearance; being a high hill full of white spots, the first of any note from the *Neversinks* in the Jerseys, to this place, and is a remarkable land-mark. At the mouth of the river is a bay, into which runs a rivulet from the south; Mr. *De Brahm* has honoured this with the name of *Grenville River*, on account of a tract of land here laid out for that gentleman, on one of the most unaccountable pieces of white sand I ever saw; which by reason of its being covered with a large growth of all sorts of trees, indicating a fine soil, I have always looked upon in the light of a natural curiosity.

From this mouth of the river southward the sound is cut into three branches, by means of two peninsulas of mangroves, divided from the main island by these lagoons; the branch which disembogues itself at *Hobé* is shallow, and full of oyster banks, about fourteen miles long; however, a small schooner, drawing five feet water, was by our people brought through here and out at *Hobé*. This inlet was shut for many years before 1769, but I have since seen it open till 1773, our people have been encamped on the same spot where now the water allows egress and regress to such a craft as the above-mentioned schooner, just sufficient to pass it: This, I suppose, has been owing to a less quantity of water coming down *St. Lucia* river for some years, because the Spaniards informed me of its having been open before. In short, this part of the country is such a curiosity, that I have many times lamented the want of leisure, and means to explore it thoroughly. My journal from the bay of *Tampe*, over land to *Augustine*, threw indeed a great light on this; but to my irretrievable loss I have missed it these four or five years: I can, however, remark upon the authority of that journal, that the fertile land is found in less proportion, in the interior part of the peninsula, in this southern, than in the northern division;—that the ground becomes more stony as we approach the meridional regions; that it is likewise interspersed with the same kind of romantic ponds, or lakes, as described on page 274,—that in the river *Manatee* is considerable fall of rocks fourteen miles from its mouth;—that above these falls the banks are very steep; —that this steepness causes the water to rise about fifty feet above its ordinary surface; a circumstance which this short river has in common with all torrents;—and that the bay of *Tampe*, into which this river falls, is the most proper place in all America, south of Halifax, for the rendezvous of a large fleet of heavy ships; the country all around being plentifully timbered and watered; the soil is poor, some of the islands along the west coast excepted, and both this and Charlotte harbour are excellent situations for establishing fisheries.

From Hobé to the latitude 25: 44, the coast is all double land, or narrow necks between the sea; there are some rivers and lagoons; on the banks of such of them as are fresh we meet with great bodies of marsh land, which may be improveable. About fifty miles north of the southern point of the main land the coast changes its course from S. S. E. to directly south; and at the head land, occasioned by this, is a large hard blue rock on the beach, out of which a large stream of very fine fresh water issues, gushing directly into the ocean; there are four little inlets between this rock and latitude 25: 35; one of these not always open; the last in the north end of the first island, whose south end *De Brahm* has thought proper to call Cape Florida, although it is by no means a cape, or head land at all, West from this is the river *Rattones*, being a fine stream, and pretty considerable, with a little good rich soil on its banks, where many tropical plants grow; at its mouth are the remains of a savage settlement. To the southward of this river is a large body of marsh, through which several rivulets of fine water empty themselves into the sound, back of the keys, which begin here, that a man may here stand with one foot in fresh, and the other in salt water; nay when the tide is out fresh water boils up through the sand. From this river and marsh the remainder of the land is a heap of stones and rocks, very sharp, and little water to be found, there being only a few ponds, and these dry in a dry season. The only growth here is shubby pine. At Sandy point, the southern extremity of the peninsula, are large old fields, being the lands formerly planted by the *Coloosa* savages; in latitude 25: 20, is a salt lake, and a remarkable *isthmus;* joining what was formerly called *Cayo largo*, or long key to the main: our researches for a passage, west of the keys, have convinced us of its being fast to the main land. In the bay of Juan Ponce de Leon, in the west side of the land, we meet with innumerable small islands, and several fresh streams: the land in general is drowned mangrove swamp. On the banks of these streams we meet with some hills of rich soil, and on every one of those the evident marks of their having been formerly cultivated by the savages. I went up into some high trees on these hills, to see if I could not spy the pine land from their tops, with a view, if I saw it any where near, to penetrate to it, through the swamps, but the nearest I saw was, by computations, twelve miles off.

These hills, among this dreary mangrove land, have apparently been the last retreats, and skulking places, of the *Coloosa* savages, when their more potent neighbours, the *Creeks*, drove them off the continent. *Punta largo*, or cape *Roman*, in latitude 25. 43, on the westside, terminates with this great bay, and the situation of the country will be best learned by consulting my charts. From this place to latitude 26 : 30 are many inconsiderable inlets, all carefully laid down in the chart; here is *Carlos Bay*, and the *Coloosa Hatcha*, or *Coloosa* river, with the island *San Ybell*, where we find the southern entrance of *Charlotte* harbour abovenamed: Let it suffice to refer my readers in general to my maps, after I shall have told him that all this point is sandy pine land, and that deer, turkeys, oysters, clams, and fish abound here surprisingly. The keys, or Martyrs, and the reef will likewise be best known by inspection of the charts: these are a heap of rocks, very few small spots on them being cultivable; *Matacombé* alone would be worth attention for a settlement; all their productions are tropical, not an oak to be found on any one, and pine trees on one only; but this reef and keys may be rendered serviceable in time of war, to any people who are well acquainted with them; the reef begins in latitude 25: 34, and the channel, between it and the islands, will admit a vessel drawing sixteen feet water; for a good way in, and at key *Biscay*, is a good place for careening craft of ten feet draught: there is good water on it, and if it should fail, the rivulets in the grand marsh will supply any quantity for a ship of considerable force, and her tender might here find the best station, for a cruize, of any I know, for there being no more than fifteen leagues from the reef to the *Beminis*, where there is likewise water, and on both shores plenty of fish and turtle, she may lay on either side in safety. The tender finds a harbour on each side, of ten feet at least; on the Florida side, the ship laying within the reef, I need not tell the seamen that she lay safe; thus cruizing across here, or one being on the station on one side, and the other on the other, it would be next to an impossibility for a ship to escape them. Of what consequence this is, in a place through which the Spaniards are obliged to send all their treasures, every one may judge. At *Cayo Tabona*, a large ship, even of 60 guns, may ride just within the reef, and her tender can always supply her with water, either from *Matacombé*, or the above-named marsh, and key *Biskay*. Few vessels can come through the gulph without coming in sight of this place, and it is generally the first land made by every sail of shipping after they leave the *Cuba* shore, Sound-Point, or

Cape-Florida, being just N. of it. *Matacombé* is another good station for small vessels, which may run in here, and the *Bahama* bank affords anchorage in case one would chuse to cruize across. *Cayo Huiso*, commonly called Key West, is another good station for a small frigate, but not so advantageous as the others, nor is there so much safety here.

Nothing further of note occurs at these islands, except that at *Cayos Vacos*, and *Cayo Huiso*, we see the remains of some savage habitations, built, or rather piled up of stones; these were the last refuges of the *Caloosa* nation; but even here the water did not protect them against the inroads from the Creeks, and in 1763 the remnant of this people, consisting of about eighty families, left this last possession of their native land, and went to the Havannah. They were a tribe of excellent fishermen; this nation was strenuously engaged in the Spanish interest, they were governed in a manner somewhat monarchical, and while they lived on these keys, were the dread and terror of the seamen; who, not being able to steer clear of the dangerous reefs which are here, escaped one kind of cruel death, to run into the jaws of one still more terrible: the inhumanity committed by the *Coloosas*, on shipwrecked mariners, is shocking even to Barbarians. A little key lying before *Matacombé* is a dreadful monument of this, it is called the *Matança*, (i.e.) slaughter, from the murder of near four hundred wretched Frenchmen, who, being cast away, fell into the hands of these monsters; who, after keeping them in the adjacent islands for some time, carried them all to this little key, which now serves them for one common grave. The people from Providence, who came here for turtle or Mahogany wood, came always armed, and had frequent brushes with them, so that the dislodging of these fierce savages has been of service to navigation. The unhappy sufferer by wreck, who escapes with life, may now be sure of safety on the shore of these islands. Having mentioned the cutting of mahogany it may be proper to observe that little or none now remains here.

I have now given as ample a description of East Florida, as the nature of this work will allow, and after having made some remarks on two curious productions of that profound and speculative philosopher, *William Gerrard De Brahm*, Esq; his Majesty's surveyor general of the southern district of North-America, I shall say something of the western province.

The first of these consists of some observations made by this singular genius, and transmitted to the *Honourable Board of Trade;* such as they are, Dr. Stork has given them to the publick, as to be depended upon.

First he says, "in latitude 27 the plantable land is scarce, except the mangrove swamps, for the cultivation of *Barilla.*" What reader, seeing this, would not imagine that these swamps were improveable, but they are a soil drowned by the sea to the depth of three or four feet; often only sand banks, and as the roots of trees grow in arches, above the ground, they are for the most part impenetrable, and some totally inaccessible to either man or beast.

2. "The trees and shrubs — are the *arboreous grape vine*, and *spice bark trees*, the *Hiccora*, *Plumb* and *Papao.*" I shall inform my botanical reader, that this *arboreous grape vine* is the *Coccolaba*, or grape tree, well known in the West-Indies, which has no manner of affinity with the *vitis* or vine, the spice bark trees are no other than two species of *Lauri*, one of them the *Laurus Borbonia*, or Red Bay, having its aromatic juices more concentrated in this climate than further north, and through the influence of the sun it is no more than a mere shrub.

The *Hiccora:* Anno 1772 he has learned to call this by its right name, *Hickorey;* but what shall we say, when it is not possible to find a single plant of this kind in the southern part of the province; nor is there any thing that looks like a plumb, except the *Chrysobalanus*, *Icaco*, or Co-co-plumb, not at all alike in kind to the *pruni* which one would judge to be meant by the plumb. The *papao* I guess is the *Carica*, vulgarly called *Papaw*, of which we find a good number.

3. "From latitude 26: 40 to 27, *there* is a branch of Hillsborough river, terminating in fresh water marsh, the principal river departing southward."

Compare this with my description of the sound into which *St. Lucia* river falls.

What a piece of accuracy do we discover in the numbers 4693, 9386, 24, 300, 950, 2800, and 37961 acres! this is of the same stamp with the matchless exactness of latitudes 25d. 48m. 33s. 26d. 53m. 35s. 24d. 57m 6s. &c. in his map prefixed to that curious publication, the Atlantic Pilot.

4. "The west side—may do in time for the cultivation of the *Opuntia* plant." When we know, that this identical plant is a weed

and meer nuisance in the very spot here mentioned; pray what is meant by a hope of its becoming a cultivated article in the soil where it grows spontaneously in great abundance?

5. "The cape and sea-coast—whose luxurious plants are the pomegranate, &c."

Not a pomegranate in all the country, except what is cultivated in the gardens at and near *St. Augustine*.

6. "The main on the west of Cape River appears to be all high land, and is chiefly covered with cedar, oak, mulberry, and gum." No such river as Cape river known to any but this extraordinary inventor himself, nor is there a sprig of any of the above plants found within many miles of the cape.

7. "In latitude 25: 35, the main upon a due west line is a mile across, and there appears a river *four* miles over, which comes either from Tampe Bay, St. John's river, or is the mouth of Hillsborough river, which in latitude 26: 50 takes a S. W. departure." Every letter of his is a forgery of the brain of this lunatic writer, no river is found or near this latitude, but *Rio Rattones* above described; what a pity it is, that the conjectures or wishes of such dreamers did not sometimes become real! Either of the communications there mentioned would open a most beneficial addition to the ease and convenience of our navigation.

8. "No fish in the white waters round the cape, at least there were none on May 13 and 29, nor is any other animal species there except sea birds, and the track of only one bear was observed." Every body that is acquainted with the immense variety and quantity of fish found here, will naturally imagine, that the fish were retired on May 13 and 29, to some general council or meeting of the finny nations, and the gentlemen from providence, who come sometimes here for the diversion of hunting a species of deer peculiar to these islands, and very numerous on them, can witness for me how true this last assertion is; not to mention the bald pate, and small American turtle dove, the red bird, the stare and American fieldfare, nor the racoon, which seems here to be an universal inhabitant in vast numbers, nor the amphibious crocodile and turtle.

9. "No sign of winter effect is visible, nor any shrub or tree species of those in the northern climate, &c. &c."

How well does this agree with the trees mentioned in the sixth article, and the impudent faseshood of the *pomegranate* is here a second time introduced with an "of which I had a full proof by the *pomegranate,* of which the trees are full of blossoms with half ripe and full ripe fruit."

10. "At Shark's-tail, Middle-river, and the head of Cape-river, are a few live oaks," and in the next article we find "hickory, live oak, mulberry, smooth bark yellow pine." The very choice for timber!

11. This agrees pretty well with the assertion in No. 9. — The conclusion of these fine remarks, is a recommendation of laying the land out in *large tracts,* the very thing which has proved the means of keeping the country uncultivated.

Can we any longer be surprized at the little, not to say perverse knowledge the people of the mother-country have of America, when we find principal officers misrepresenting facts to the chief rulers of the land? I make no doubt but many good folks, in other offices, have much misrepresented facts of another kind, as this man has done the description of the affairs in the department he was intrusted with.

But if this narrative deserves contempt, and to be exposed, the latter production of 1772, with its very improper title of Atlantic Pilot, evidently bearing marks of insanity, demands our pity; here we see an account of an unnatural change in the face of the country, which for many reasons never could have happened but in the brain of this Bedlamite, from whence also seems to originate the name of *Tegesta;* he turns one peninsula into broken islands, another into sunken rocks; what a havock of jumbling this Hercules makes! when in this unmeaning *chaos* he joins and disjoins, turns water into land, and land into water, calls the current from Baffin's frozen bay, to join with the velocious stream of Torrid Mexico and Florida; and again makes them form a *vortex* reverting back to their points of departure; magnifies a parcel of pitiful flats into a gulph, with the sonorous name of Sandwich, or into a lake with the *pretorian* title of grand affixed to it, and all this does not cost him more trouble than a few strokes of his inimitable pen; nay more, he metamorphoses, the hunting seat of the Prince of Orange into a wolf, and turns a woman just delivered from child bed travail, and her child, into a paradise*.

---

*Key *Loo,* so called from the *Loo* frigate which was cast away on it, he has changed into *Loup,* the French name for a *wolf,* and the two keys by the English called Soldier Keys, from the multitude of those animals found on and near them, are by the Spaniards titled *La Parida y Su liguela,* but this curious raver has called them *Los Paradizos.*

Not to say much of his arrogant false claim of his having discovered the navigation through the chennel, inside of the reef, which numbers of people, even Englishmen, knew better forty years ago, than it is likely he ever will know it; I cannot forbear mentioning the alteration of well-known names in a place of so much danger; can the arbitrary imposition of the names of *Dartmouth, Littleton, Pownal, Hawke, Egmont, Huntingdon, Holbourn, Keppel, Fox, Townshend, Ellis, Oglethorp, Reynolds, Dyson,* and about a legion more, of inferior great and little men, make up for the mistakes they may occasion? Or can these statesmen and heroes owe him thanks for the confusion occasioned by the jumbling of their names, like dice in a box? The formation of a reef, "out of sight of any land," west of the dry Tortugas, is likewise a child of his own brain; and we are apt to think him finding fault with the goodness of the Creator, in not making shoals enough, for there is no such bank any where west of the dry Tortugas.

But his placing soundings in his draught deeper by three feet than they really are; and his advising people who intend to go through the Gulph of Florida, to take their departure at the Havannah, and steer due north, in order to make, what he calls, Cape Florida, seems as if calculated on purpose to destroy ship, goods and people: happy is it for me that our present navigators know the navigation so well, and for the benefit of trade I hope his pamphlet will never serve as a guide to any man that is a novice, and chances to come this way.

As a specimen of his marine knowledge, observe him telling us, that at times of a westerly wind, the Atlantic coast is the most eligible lee for navigators who do not chuse to take the stream; as he has just before defined this Atlantic coast to be the Florida shore, it must consequently at such seasons be the weather shore; but cease to wonder at this, when I tell you, that the method of his own, which he refers to, for taking the variation, is a mechanical one and very tedious, and that he did not know how to do it astronomically, by an amplitude of the sun, till I taught him in 1769. I am almost weary of removing all this dirt and would leave the man and his performance, but my love for truth in natural history obliges me to trace him, even through his presumptuous rummaging of the kingdoms of nature, of which he knows as little as of nautical affairs.

To support the ridiculous hypothesis, he has broached about the change of his ancient *Tegesta,* he says, amongst other arguments, in page 9, "myself and people, employed by me in this service" (of which number, I the present writer was one) "have these three years observed many places where fresh encroachments appear to this effect: even the vast quantity of scattered large old trees, washed out with their roots on all shores of the islands, and out in the shallow sea, between the islands and the main, testify, that they lay on the spot of the former continent and peninsulas, where their genus and species formerly flourished." I came through this place four years before I knew this man, and from that time down to last year I have seen many changes and alterations in the positions of these trees, some of which come down the rivers in Cuba, in land-floods, but more down the Mississippi; and by the currents are here forced on the reefs and keys.*

A few of those from Cuba being tropical, might serve to support the reasoning of the Atlantic Pilot, but many kinds never grew in Florida; the coco-nuts found on the shore likewise convince us, that Cuba sends much of her outcast this way; but when we consider that nine out of ten of the trees thus found are oak, cypress, red cedar, elms, and above all, beach of the greatest magnitude (none of which ever grew in this climate) it is done with *Tegesta,* she is fallen never to rise again!

He has however this time forgot the *pomegranate;* but how do we know what is meant by *Papaios,* distinguished from *Papaw;* some new genus in botany, no doubt, has the Spanish name of the Papaw bestowed on it by this excellent naturalist. Observe him last of all turn systematist, when he tells us "that a species of prawn (shrimps growing to the size of five pounds" (he might have said fifteen "are improperly called lobster;" ô *Philosophus eximius!* The lobster which is the first in rank among the families of *Astaci,* must here give way to the most diminutive of the whole genus, which is placed first, and that forsooth because in *Mr. De Brahm's* opinion there can be no lobsters without claws:—but enough of this, let us proceed to the *topography* of West Florida.

The river *Apalachicola* is the boundary of the two provinces; before its mouth eastward, and a little westward, we meet with St.

---
*Let us for a moment calculate, how long the worms will leave such a tree as a monument, and we shall find, that his change of *Tegasta,* at least in part, must have happened of late years.

George's Islands, well known for the sufferings of *Pierre Viaud*, &c. after shipwreck here. I must remark, that the relation we have of this affair is a great exaggeration of facts; people in that situation may suffer, but on a place like this, where plenty of fish, crabs and oysters, are to be had, as well as water, for the trouble of digging, their sufferings cannot be great. Our pity is much lessened upon finding a reasonable man so debilitated by fright and despair, that he cannot make use of a plenty provided by providence. All along this coast fish are in such abundance, that both myself and people, when not inclined to fish with a hook and line, have struck many hundreds with a stick sharpened at one end, and hardened in the fire, in lieu of a harpoon; I will not mention much about the exaggeration of distress, which runs through the whole of this performance; but the story of Madame *La Couture* bringing a turkey and her eggs, out of the woods, at the time when relief had reached her fellow sufferers, appeared to me too palpable a falsehood to be credited, because these birds were never found on any island from Carolina to this place, except formerly on Amelia island; (which is separated from the main only by a little creek, in some places scarce ten feet across) but supposing one to have been accidentally here, I will readily allow that the eggs may have been found, and taken by Madame *La Couture;* yet as these birds are the most shy animal we meet with, (insomuch that a deer, in places where he is frequently disturbed, is incomparably more easy to come at than turkies) certainly she must have been a more expert huntress than ever I heard of before, and I should be after all very much inclined to think that the poor woman, and her companions, had mistaken the Mexican vulture for a turkey; but that from the relation we naturally judge there were many eggs in the nest, which oversets that kind of reconciliation, wherefore, finding it was at the time of their relief, I went to Mr. *Simpson,* (the interpreter for the savages, who was one of the people in the boat at the time of affording help to those unhappy people, and is a man of veracity) and asked him if it was true that the woman brought in a turkey, and her nest, of which he and his companios shared; he told me that it was no other than a crab of the kind called in the southern province a king crab, and to the northward a horse-shoe, but he saw no eggs of any kind. As this fish bears a distant resemblance to a turtel, perhaps Mr. *Viaud* took it for such, and wrote *tortue*, which may have been translated turtle too, but by a mistake common enough in printing offices, the compositor may have made it a turkey*, and the corrector, either through inadvertency, or not considering how largely he launched into the marvellous left it so.

But to proceed, a little west of these islands we find Cape Blaze in latuitude 29: 47, and the Bay of St. Joseph, to the N. W. of which is St. Andrew's Bay, and W. by N. from here, *Santa Rosa* bay and islands; in the east end of this bay is *Chatto Hatcha;* which last running E. and W. makes the eastern shore of Pensacola entrance. All these places are very arid, dry sand, near the coast; farther in towards the Upper Creek country, which is in some measure described in page 91, it is scarcely known, but what I have seen of it is much like Georgia, and the north part of East Florida.

Pensacola has about an hundred and eighty houses in it, built in general in a good taste, but of timber: the town is laid out in an oblong square, near the foot of an hill, called *Gage-hill;* and by means of two rivulets of excellent water, which almost surround it, is the best watered of any I know on the continent. The harbour is spacious, and here are three considerable rivers, viz: *Escambé, Chester,* and *Middle* rivers, besides several lagoons and rivulets of no note. *Chester-river* is full of islands, great and small, abounding principally with the *Cupressus Thyoides. Middle-river* communicates with several lakes; the water of this river is so cold, that the savages call it *Weewa Cosupka,* which in the Creek language signifies cold water. The *Escambé* is navigable a considerable way up, but rapid, and about twenty miles from Pensacola we begin to meet with some spots of fertile land, variously timbered. Twenty-eight miles from the town, and on the banks of this river, on an eminence, are the remains of a Spanish out-guard, or stocado fort; about seventy-eight miles from town, this river forks, a branch called *Weewa Oka,* or *Weeoka,* by us little *Scambé,* coming from the N. W. runs into the main river; this branch where it is crossed is about sixty feet wide, and the ford is a bottom of small gravel, which is not common here; the road that crosses it leads from Pensacola to the Upper Creek nation; the distance is two hundred and ten miles; and the heads of these rivers are

*I find mention made also of a *bustard*, it is well known that there is not one of this kind of bird in all America, however a *bustard* is in French called *eutarde*, and this is the name given by the French to the blackhead Canada goose, which abounds here in the season, when *Viaud* was on this coast, this the translator seems not have been informed of. I perceived many symptoms of an imperfect knowledge of the French language, by the translator, however, this is not a place to point them out, only out of complaisance to the sex, I will inform her that *casse tete* means a *tomahawk*.

all within a few miles of the savage towns. Fourteen miles farther west from Pensacola, is the *Rio Perdido*, or lost river, of inconsiderable length and navigation; and about forty miles west of this, we meet with *Mobile, Tombechbé* and *Taënsa* bay and rivers; into which last the *Aibama*, or *Coosa* river falls. All the country between this last river and *Chester-river*, and probably all to *Apalachicola* is very favourable for stocks of cattle; many noble cane branches, mixed in the pine land, affording them excellent food, and they multiply very rapidly. Lovers of minute geography, by examining my maps, will find that the lower part of the coast between *Mobile* bay and *Perdido*, and so on to *Pensacola*, may easily be made navigable for boats in land, there being already a rolling road out of the river *Bon secour*, into a creek running into *Perdido;* but this being mere matter of speculation we must leave the practical part to some future generation. The country west from *Mobile* bay, to *Nita Albany*, or Bearcamp, at lake *Maurepas*, is much of the same kind as that already described; only we find more fertile planting ground in it. The timber here is nearly the same as in the north part of East-Florida; but it is worthy of remark, that not one of the *Borassus*, or Cabbage tree, is found along this whole extent from *Apalachia* westward; which does not agree well with a report in a letter to a certain *George Lookup*, Esq; in *Robinson's* history of *Florida*, where he says, that the cabbage trees along this coast, rear their lofty heads above all others; but so curious a piece as this pamphlet is ought not to be ransaked, I will therefore proceed to give some extracts from my journals through the western parts of West Florida, as being the best way to give my reader a true idea of this country.

In the following pages the distances are accurately measured, by such methods as circumstances permitted.

On Saturday the 20th September, 1771, at 4 P. M. I left *Mobile*, and encamped that night in pine land, near a spring to the north of the path, and at the foot of a hill, six miles from town, to the west a little by south.

21st *Opaya Mingo*, a Chicasaw warrior, of our train, was this morning sick, on which occasion I saw one of his companions cut his temples with a flint, and applying a cane about four inches long, to the scarification, suck it till he nearly filled it with blood, then threw it out, and repeated it several times; this is something like cupping: we were obliged to leave these two behind, and proceeded this day chiefly to the N. W. through pine land, and encamped this night at the head of a branch of Dog-river, fifteen miles and three quarters from town.

22d. Crossed several creeks, and headed others, running all eastward; we advanced twenty miles and a half through pine land, and encamped near a creek called pine logg; in the afternoon we came to the first considerable ascent, where the ground is gravelly, here the road goes off to *Yoani*.

23d. This morning we went over a very narrow ridge, leaving Dog-river on the left, and Tombechbé on the right of us. Crossed several heads of cane branches, running different ways; saw many grass ponds, and Indian camps, and crossed a large cane branch, called by the Chactaws *Coosak Hattak falaya;* proceeded eleven miles and encamped on the waters of *Pasca Oocoloo* river; here saw a good deal of chestnut and other timber, besides pine trees.

24th. This morning we came past some good low grounds, and saw the first oaks of the kind called black jacks: crossed several creeks, and went through some swamps and oak land; encamped near a branch, and here saw the hieroglyphick No. 1, in page 108, advanced this day twenty-one miles.

25th. This morning we had a great deal of rain, which continued by showers all day; the two savages left behind joined us. We advanced but five miles and a half; encamped this night about two miles W. S. Westward from the path, near a lake abounding in excellent fish, particularly pikes, perch, and red eyed chubs, of a very superior size.

26th. Being rainy we lay by, and recruited our provision with abundance of fish, two deer, and two turkies.

27th. Went over a good deal of gravelly ground, mostly pine land, and encamped at *Bogue Hooma*, (i.e. Red Creek, our boundary with the Chactaws; here we saw the first rocks on the south side of the creek: we travelled only eleven miles and quarter; the creek being very high we were obliged to spend a great while in ferrying our goods, and swimming our horses over; the stream runs westward;— shot three turkies.

28th. After six miles travelling, we came to oak-land, being short hills, and all day we found the land pretty much mixed;—crossed many springs and rivulets;—saw the head of a savage stuck on a pole, with many other marks of our being on the theatre of war; we left the road to Chicasawhay on the left;—all this afternoon we leave a considerable stream on our right, running N. eastward:—advanced this day fifteen miles and a quarter, the last three or four miles the land was chiefly oak-land: at night we encamped at a war-camp, near a branch running into the above stream, from which we were one and an half mile to the west, using the ordinary precautions*.

29th. Ascended and descended several considerably steep hills; found the land mixed, chiefly oak; at three o'clock went through the savannah called *Poos coos Paähaw*, about one mile and an half across, with here and there some points of wood land on it; and again crossed some hilly oak land; and after thirteen miles travel encamped near a branch of *Pasca Oocoloo* river; here a very plain path goes eastward.

N. B. The water in the savannah has a taste of lime.

30th. Almost as soon as we left camp we crossed a branch of *Bogue aithee Tanné*, vulgarly *Bakkatané*, having the low land of this river constantly in sight, about half a mile to our left; travelled chiefly through pine land, and some *hurricane ground*†, and after journeying eleven miles and three quarters, we encamped at *Hoopab Ullah*, (i.e.) the noisy owl, where we saw the hieroglyphick No. 2, page 102.

1st October, went through a vast variety of hilly, stony, boggy, swampy and oak land; passed some very elevated hills, and crossed several rivulets: encamped at *Oku Ullah* (i.e.) the noisy water, a creek; having gone twenty-two miles and a half this day, we passed by three graves within the space of three quarters of a mile from each other, the first of a solider of a detachment that went to fort *Tombechbe*, the second of a savage, who went by the name of rum-drinker, after whom the hill is called rum drinker's hill; and the third of one Mr. Brown, a very considerable trader in the Chactaw and Chicasaw nations.

2d. At four miles distance from our last night's encampment we crossed *Pancha Waya*, the last water running to the south-west, and two miles further we left the path going to *Coosa* on our left. Our road all this day led through the same kind of rugged, uneven, stony, gravelly and swampy ground as yesterday, as also a hurricane ground

---
*See page (129).
†Tracts of wood formerly destroyed by hurricanes are so called.

and scrub hill; some of the stones we saw were very large, and the hills considerably elevated: having travelled eighteen miles and three quarters, we encamped a quarter of a mile west of a Chactaw village called *Paonte*, near a delightful spot but deserted on account of the war; here we got some peaches, plumbs, and grapes; six miles W. S. W. from the camp is the Coosa town, of Chactaw. A showery day.

3d. Very heavy rain from midnight till noon, at two miles and an half from our camp left the road going to *Haänka Ullah* (i.e.) the bawling goose, a chactaw town, to the left; travelled only four miles and an half through mixed but pretty even land, and encamped near a branch of *Sook han Hatcha* river, the streams being very high the ferriage delayed us much.

4th. This day we met some traders and savages, and travelled through several deserted fields, and one deserted town, called *Sapa-Pesah:* the ground is generally very uneven;—here we crossed several paths, and were joined by three Chactaw savages, who extricated us from a great difficulty, as we found no convenient logs to ferry us over a large branch of *Sook hanatcha,* till they shewed us a place half a mile down the stream; and for a small consideration ferried our goods over, while we swam our horses across: having traveled ten miles and three quarters, we encamped on the north bank of this stream, called *Hatchatipke.*

5th. We came after half a mile's journey to a deserted town, called *Etuck Chukké* (i.e.) *Blue Wood*. This morning, at day-break, we heard the report of fire arms, at a little distance, which, being very often repeated, we soon guessed it was what we found it to be a little while after; that is, an action; the savages always attacking either at day-break, or at sun-set; crossed many paths and fields, and after ten miles and an half journey, we found one of the victorious Chactaws presenting us the fresh scalp of one of his enemies. Here were the first inhabited houses, being a kind of suburb to the town of east *Abecka;* here we crossed the branch of *Sookhanatcha* river, which gives name to all the rest, and at a mile and an half further, we came to the town itself, and put up at the house of an old trader, called *Hewitt,* and were civilly treated by both savages and whites. Of the first we found here three chiefs, not a little exulting in their late victory, and the women in every part of the town were performing their savage *orgies* and dances; we staid here till the 9th, when part of my company proceeded to the Chicasaw nation.

9th. Agreed with a very worthy young man, one Mr. *George Dow*, to serve me as a guide through the Chactaw nation; and that day we went through *Ebeetap-Oocoola,* a town on a hill, where the savages have a large stockade fort in their own manner. Such there are also at *Abeika*, which two places are the frontiers against the Creeks. We went this day south westward, through *Chooka-hoola*, part of *Oka-hoolah*, through *Hoola-tassa*, which last is four miles from *Abeika;* then nearly south three miles more to the ridge of east *Moka-Lassa*, then about a mile and an half westward to *Ebeetap-oocoolo-Cho*, where we stopt at the house of one *Foster*, a trader; then went nearly S. E. for two miles, and came to the middle of east *Moka Lassa;* from here continued S. E. a little more than two miles, and entered *Haänka-Ullah;* still went on the same course for a mile and a quarter, and then were at the middle of said town, and refreshed ourselves for an hour. The town of *Oka Loosa* lays S. E. three miles from here; then returned, in a round about manner, through another road, back to Mr. *Foster's* house, where we slept.

10th. Proceeded first W. N. W. about two miles, afterwards W. S. W. for a mile and a quarter; this brought us into *Okaattakkala,* a neat little place for a town of savages; went about a quarter of a mile into it, and put up at the house of Mr. *Dow*, my guide. About three miles W. S. W. from this are the ruins of *East Congeata*, destroyed in the civil war of these people, by the western party. Hitherto travelled chiefly through corn-fields; we lay here till the 21st, partly to refresh our horses, and partly on account of a violent fever, which I was troubled with; during this time agreed with a savage to join us as a guide to the S. W. of the nation, his name was *Pooscoos-Mingo*, or king of the children.

21st. Went S. westward, a mile and an half into *Yanatoe*, crossing many paths, and several old, as well as some cultivated fields, which lay interspersed in the woods; crossed *Bague-fooka*, and the heads of *Poreetamogue*, with several other creeks, and after riding something better than seven miles and an half, mostly S. W. we ascended a long hill, steep in its first rising, and when on its top we had the view of a high ridge, to N. W. with a deep valley between them. Went south, and after a mile and a half crossed *Bogue Chitto*, a large creek, which runs S. E. into the river *Bogue-aithé-Tannè*, in three quarters of a mile more went through the deserted old field of *Coosak Baloagtaw;* about a mile and a half further we crossed a creek of that name, emptying into *Bogue-Chitto*: a quarter of a mile further we crossed a large pond upon a beaver dam, and in half a mile more stopped to bait, at a creek called *Pooscoos te Kalè*, in a deserted village of that name. At near three o'clock proceded S. W. by W. after a little more than two miles, crossed a deserted field called *Pooscoos tekale, Hoca;* two miles and three quarters further crossed *Tallé Katta,* a large branch of *Chicasaw-hay* river, then came into pine land, for the first time of these fourteen days: went south, a little westerly, and came to camp at a place, where we found some old camps convenient, near the head of a branch of *Talle Hatta;* having this day travelled twenty-two miles and an half, all in woods.

22nd. We travelled through various kinds of wood, and something of uneven ground, chiefly to south; and at twelve miles from our camp we crossed a river near eighty feet wide, here said to be the principal branch of *Chicasawhay* river; and is called *Aitheesuka;* the ford is a flat rock, the land on each side here is pine land, and rises high above the bed of the river: continue in pine land four miles and a half more, then come into low land, cross some creeks and savannahs, and after having travelled twelve miles more, chiefly S. E. from the river, I came to camp on the side of a pine hill. My weakness not suffering me to go further, I ordered our people forward to *Chicasawhay,* to provide necessaries, except one lad, whom I kept to accompany me.

23rd. South eastward through some hurricane and other uneven ground for five miles and an half, then went through intervale for a mile and a half, at the end of this we go up a considerable hill, from whose top we see a very high blue ridge to E. and S. E. a great way from us, and at the end of nine miles and three quarters we arrived at *Chicasawhay,* and stayed at the house of *Ben James.*

24th and 25th. Employed in going to and coming from *Yoani.* The road that leads to this place is chiefly pine and shrubby oak land: at five miles from *Chicasawhay,* a middling branch of *Chicasawhay* is crossed: it is called *Owhan lowy;* judging this sufficent to give an idea of the face of the country here, I shall only subjoin that on the 7th November, having made a circuit of the nation, we returned to *Hewitt's h*ouse at *Abeeka,* our road from *Chicasawhay,* leading chiefly through fine improved, and many among them rich fields, and a large number of considerable towns and villages of the savages.

10th November, having dismissed *Pooscoos-Mingo*, we went to the *Chicasaw* nation, through a road leading in general over stiff clay land; saw very little else but white oak, and that no where tall, occasioned by the stiffness of the land; crossed only two rivers of note, one *Nashooba*, the other *Oka tebbee haw;* no remarkable ascent or descent on the whole road; crossed many savannahs, the distance is about an hundred and fifty-nine miles from *Abeeka*, to the *Chicasaw* towns; (described in page (60) where we arrived on the 18th November, and put up at the house of one *Buckles*, a trader: our course hitherto has chiefly been north. Here I stayed till the 8th December, and observed this house to lie in nearly 35° north latitude.

I cannot help relating in this place an anecdote of Mr. Commissary in this nation: this Gentleman had engaged to me, and promised Mr. *Stuart*, that I should have all the necessary assistance I might want; I went to his plantation, which lays at *Paön titack*, eleven miles from the place where he ought to reside, and to my astonishment I was here treated worse than in any place I had been at: so far from providing horses, which was one of the articles I wanted, I was obliged to give him one back, which I had borrowed at the Chactaw, in lieu of one that failed me. I saw clearly, that I was an unwelcome guest in every respect, however, I procured from him a recommendatory note, to the above *Buckles*, and he told me that *Buckles* knew of two canoes, the choice of which I might have for my intended voyage, down the river; but when I spoke to Mr. *Buckles* on the subject, he denied ever having spoke about any canoes, nor did he know of any, and as to the note, I might have any thing I wanted on any own credit, by giving an order on Mr. *Stuart*; but with the commissary he had nothing to do. Thus I was obliged to do all myself at last; this is in some measure an instance of the inutility of the commissary's office; I sent Mr. *Dow* with three negroes to the river, to make a canoe. During my stay here, on the first December, it began to freeeze very hard, on the fourth the ice in town-creek was four or five inches thick; and on the 6th in the morning we could scarce keep water from freezing in a close house, and near a good fire, but I saw no snow during all the time of my stay. Nothing was more entertaining than the surprize of the savages, at seeing me take observations of the solar altitude, the mercury I used for an artificial horizon, was a matter of great wonder to them, particularly, when I shewed them its divisibility, and the succeeding cohesion of the globules.

8th December, in the afternoon I proceeded from Buckles's house, and travelled E. by S. nearly five miles and a half; then crossed *Nohoola-inalchubba*, or the town Creek, mentioned in page (63) then went about S. S. E. for a mile and a quarter more, and encamped on a hill, near a small lake, in which we heard all night a great number of ducks. We met this day eleven savages, of the Creek nation, and three more of the Natchez, who live with the Upper Creeks, coming through the *Chicasaw* field, with the *Death Whoop*: they had surprized a hunting camp of *Chactaws*, on a branch of the river *Yaasoo*, called *Tallé Hatcha*, at a time when the hunters were absent, and took away all their skins, two horses, and five women of the *Chacchooma* tribe, belonging to west *Congeta*, in the *Chactaw* nation: they took care not to stay for the return of the men, as the camp indicated them to be numerous; but because they had not a scalp to shew, these devils incarnate had scalped the eldest of the women they carried into captivity, yet let the wretch live, and she was one of the number we saw. The *Chicasaws* were all abroad that day; but I observed by the behaviour of some superannuated fellows, that they disliked the *Creeks* behaviour of bawling the *Death Whoop* through the field; and having asked the reason, I was told, that as they were neutrals in this war, they did not like to give umbrage, which the *Chactaws* might take at this, and if their men had been at home, they would not have suffered this small victorious party to make use of this kind of triumphal entry here; thus we see these savages strictly observing a neutrality in form.

9th. Proceeded, and at 3 miles, saw a large *tumulus*, which was the only remarkable thing in this road, we went over one steep hill, and crossed three or four branches, and the last two miles and a half of the road we saw much pine mixed with the oak; having travelled nearly twenty three miles on a course S. E. by S. we arrived about four o'clock P. M. on the bank of *Tombechbé* river, here called the twenty mile creek.

10th. Examined the canoe, but found her too heavy for our purpose, therefore dispatched Mr. *Dow* (who knew the ground, and was well acquainted with the woods) to see if he could not find some canoe of the savages; he found one about noon, in the mouth of *Nahoola inalchubba*, two miles and a half west from our camp, therefore went and encamped on the bank of said creek; the opposite bank was very high steep hills, above hundred and twenty feet perpendicular, and some pines on their summits; found the canoe a very sorry one, but it being *Hopson's* choice, must venture in her.

11th. We spent in fitting our craft; since the 7th we had very bad weather here, one hour it is sharp, clear, and intensely cold, the next cloudy, sultry, with lightning, thunder, rain and hail, the wind veering round the horizon in an incredible manner, which made my situation very disagreeable.

12th. We had some tolerable weather, and finished fitting our boat, but she was so miserable a tool, that I believe few men would have ventured in her.

13th. All the company returned to the nation with the horses; kept only Mr. *Dow,* and my servant, to be the companions of this fluvial expedition; while we staid here we caught some beavers. At half an hour past one, P. M. we proceeded from our camp in *Nahoola Inalchubba,* or Town Creek, and in twenty minutes were in the river. We passed one bluff, where the French formerly had a fortified trading house, about one mile below the mouth of the creek, on the west bank; we passed one creek on each side, and left three small gravelly islands on our left; for the rest of the land was low on both sides;—having come five miles and an half we came to camp on a sand bar at half an hour past three, P. M. the river pretty rapid, and found only one log, which detained as a quarter of an hour.

14th. Proceeded at ten o'clock, A. M. in half an hour past a bluff on the east side, and in half an hour more found a very tall cypress laying across the river, by which we were detained an hour; but cutting the top away we passed on; about three quarters of an hour after had some pine land* on the west bank; in general the land on both sides is low. About two o'clock P. M. past a small island; about half an hour after two the bank on the west side was about seventy feet high; // continued between pretty high banks till half an hour after three, when we encamped on a high sand bar, on the west side of the river, current unequal, and the obstructions of no note, since we cut the tree; travelled this day nine miles.

---

*When the pine is mentioned hereabouts, I mean the *Pinus Abies Virginiana, Conis parvis subretundis,* or the balm of *Gilead* pine.

//N.B. The river was high.

15th. Departed half an hour past ten o'clock A. M. past several islands, and found the bank, on the west side, in many places high, we saw in many others high and intervale oak land;—not so much drowned land as the former days;—passed through several rapids, and at half an hour past three, P. M. came to camp at the end of a large high pine bluff, on the east side, having travelled this day fifteen miles.

16th. At half an hour past nine o'clock, A. M. proceeded and came past a variety of high and low land; shot some ducks and teals; but unfortunately this forenoon lost our gun, being torn out of my hands by a snag of a tree, and as it sunk in deep water, it was irrecoverable, the river being rapid. About two o'clock P. M. past a short hill, on the west shore, upwards of eighty feet high above the present level of the water;—passed a few creeks and lagoons, three islands and several rapids; we came down fourteen miles and three quarters, and at half an hour past three, encamped on the east side in a hollow.

17th. Proceeded a quarter before nine A. M. and in less than half an hour saw a hill, a little inland, on the west side; and an hour after this we had the mouth of a river on the east, this is called, at its fording place, in the trading path, the last branch of *Tombechbé*; none of the banks we past this morning were very low, and on the west side, opposite this mouth of a river, the banks fifty or sixty feet higher than the present surface of the water. The weather being severely cold I was obliged to go on shore and make a fire, on this highland, half a mile below the above river, near the mouth of a stony creek, which we took to be the *Sonac Tocalè*, which we had crossed in the road going up; here we remained till noon. We passed three islands this day, and the general run of the land is middling high on both sides. The weather was very severe, which obliged us to encamp at one o'clock in the mouth of a creek, called Old Town Creek, at the foot of an hill, rising about seventy feet out of the water, having come about seven miles; it rained very hard all the afternoon, and the next two days, by which means we kept in camp till the

20th, at noon. Half a mile from our camp the high land ends, and we travelled but one hour and twenty minutes, when the severity of the weather obliged us again to encamp, having come six miles, and in that distance found chiefly low land: past four islands, and as many rapids, one of them a bad pass: our camp was on a place of fine rich, middling high land, which Mr. *Dow* reported to be very extensive. We lay here on account of bad weather; such as cold, rain and snow, till the

26th, when at a little past twelve we departed. During our stay here the water gained near four feet in height: in one hour and an half, having gone near five miles, and past six or seven islands and rapids, we were at the mouth of the river *Oca Tibehaw*, the lower road between the *Chactaws* and *Chicasaws* crosses this river about two miles above this place;—about one mile below this, having past an island, we saw a very remarkable bluff on the west side, rising about fifty feet out of the present level of the water; this is near a mile and a half long; the bluff is covered principally with *Juniper*, or cedar shrubs, and in it are two or three gullies, in which as many springs come trickling down; for about two thirds of its length a bank projects out at its foot, having a flat and very even surface, without any plants, (a little grass excepted) growing thereon. This surface projects about nine or ten feet from the bluff into the river, and was about eight feet above the level of the water, it looks as if made by art, and if placed near any town of note, I do not doubt would be much used as a walk. The bluff ends in two little hills, and a small island is at the lower end of it; all which, added to its being in the form of a crescent, makes it have a very romantic appearance. From hence we went past one more bluff, a considerable island, and a rapid; in which last we had low pine land on the east bank, and got into the mouth of a creek, called eleven mile creek, which we went up half a mile, and at three P. M. we were forced by the rain to come to camp, having travelled this day, ten miles and a half.

27th. The rain obliged us to lay by all day.

28th. At 11 o'clock we embarked, and had more rapids to go thro' than on any of the former days; likewise saw more high land than before. I think it remarkable, that we see so very few outwaterings of creeks, or rivers, into the great river. An hour and an half after leaving camp we saw a bark log, just landed on the west side, and evident marks of people have just landed; this was in a long reach, and we had seen a smoak, but when we came near the bark log, the smoak vanished all at once. We soon found these people to be a war party of Creeks, who perceiving our boat, had put out their fire, which on these occasions they make of hickory bark and other oily matters, that yield little smoak. We therefore put on our hats, which we had not on before, it being a fine agreeable day, and rather warm, and laying on the paddles, did all we could to shew that we were white people. It was fortunate for me that I had not brought a savage guide with me, which would have exposed us to a volley from those warriors; we did not see them, but we knew by the suppression of the smoak, that they had discovered us; they were undoubtedly on the top of a pretty high bank, in ambush, so we let the boat flow past them. N.B. We have discovered many of those bark logs, made of canes, both above and below this place, upon which the war party ferry over. Half a mile lower down we saw the mouth of a middling large creek, on the east side. We past four islands this day, in two hours and an half's time. We saw one of the high savannahs, which bounds here on the river, with a delightful grassy bluff: the river is here above two hundred feet wide. Having come twenty three miles, we encamped at four o'clock, on a gravel bar. We saw many places that appeared like old fields as having been formerly cultivated.

29th. At nine o'clock proceeded; the first half hour we passed chiefly through drowned lands, and then came to a large lagoon, going to N. W. in which were innumerable geese and ducks; but the west bank of the lagoon is a handsome bluff. In half an hour more we came to a creek's mouth on the west: we saw again some spots bearing marks of former cultivation, and more of drowned land; the river in general this day rapid, with many islands: at half an hour past three, P. M. having travelled thirty-five miles, we encamped on the east side, in a lagoon, on a high bank, where, for the first time, we saw the rich ground clear of large canes: this being timbered, chiefly with the shag bark hickory, iron wood and Spanish oak.

30th. This morning we had our boat loaden, but it began to rain, and thus were obliged to unload again. Here we found a canoe which was well made, but had been by the savages, on account of the war, scuttled, and rendered unfit for use, however we found, that if we could make her any ways tight, she would be more safe than the one we had, and this we effected by the help of wedges, clay and leather. Our provisions beginning to grow scant, and having lost the only gun I had taken with me, I began to be uneasy, especially as we found the beavers become less plenty. The weather was uncommonly bad till the evening of the third of January, 1772, when it cleared up, and next morning

4th January, 1772, at a quarter before 11 o'clock, A. M. we proceeded in our new craft; and by half an hour past one, P. M. we had past high banks, and two islands: at this hour we had one of the savannahs on the west side, the bank being here about four feet out of the water: at a quarter before four we were at a considerable island, having gone since the last savannah between low banks; a quarter before five o'clock came to camp under a low bluff, on the east side. The canoe proves very leaky, and on unloading we find a great deal of our bread spoiled. This day came twenty-two miles and an half, the canes and timber are here exceeding large.

5th. At a quarter past ten embarked: high banks on both sides. At half an hour after eleven, past the mouth of a river from the east; this is called by the savages *Nashebaw*. The land here is exceeding rich, the canes very large, and we saw a species *phaseolus*, in great abundance, along the banks. About a mile and a quarter below the creek, we met four savages from *Abeka*, in the *Chactaw* nation, to whom Mr. *Dow* was known. The river has risen considerably since yesterday; the current has been for these two days almost uniformly at the rate of three miles per hour. It is remarkable, that though the velocity of our way was not much above two miles per hour, independent of the current, yet we had several instances of having evidently out-run the flood at night, in so much that it would scarce reach us again before morning. We came to camp near the savages above-mentioned, having come seven miles and an half; they had a good canoe which I intended to purchase. The weather was very cold to day. Lost a silver spoon at our last camp, which Mr. *Dow* proposed to go and fetch; but he found it impracticable to cross the *Nashebaw river*. I agreed with the savages that they should hunt for us to procure provisions. At night it rained; our camp was on the east side, in a very rich spot. We have not yet seen any sand or gravel, except on the bars and islands in the river, the soil in general being clay, or loam, with a dry black mould.

6th. After a great deal of persuasion, I bought the canoe from the savages; and they brought me in two deer, and a turkey; for all which I gave them five yards of blue strouds, two powder horns, a knife, and some small shot. I described our last camp to them, and desired them to look for the abovementioned spoon, directing them to leave it in their own country, with Mr. *Dow's* partner, who lived just by their homes, and gave them a note to him, desiring him to pay them for their trouble; but when they understood it to be the *white stone* (i.e.) silver, they declined going purposely for it; but promised, if chance led them to the place to carry it as directed, for they, not being able to work it, in case the trader refused to pay them, they would lose their labour; but had it been the *fat of the earth* (i.e.) lead or pewter, in that case they might make bullets, or ear-rings of it, and then they would not take pains in vain: towards evening they left us, and during night we barbecued our venison, to preserve it.

7th. At half and hour past seven o'clock A. M. embarked in our new craft; all day we past between high banks, some steep, some sloping; several as high as eighty feet above the surface of the water; one of these has an extensive savannah on the top. At the end of nineteen miles and an half, on the west side, we saw the mouth of the river *Noxshubby*, or *Hatcha oose*, its banks are high on both sides: here seems to be the true theatre of the war, for the bark logs are very numerous. The river is widened now from two hundred to two hundred and fifty feet. A mile and a half below the mouth of *Noxshubby* is the first bluff of a red ochre—like earth, very high and steep; and about eight miles lower a white one, being a kind of stone almost as soft as chalk. Towards evening we met with several sand bars, and little isles. At four o'clock P. M. we came to camp on a low spot, on the eastside, having gone down the river forty-two miles and a half; all day dull weather, and at night rain.

8th. Proceeded half an hour after 8 A. M. and having gone a little more than a mile, we reached the mouth of the creek called *Eetomb gue bé* (i.e.) Crooked Creek, on the west side; from this creek's name the French derive their *Tombechbé*, the name of the fort which stood here, and which has again given that name to the whole river. We went about half a mile further down, under a high and steep bank of chalky stone, and arrived at the ruins of the fort, by means of which the French kept all the savages in awe. I went ashore in the old fields, and drew a view of the ruins; this is about forty miles east from the town of *Abeeka*. The river is not sixty feet wide here. About a mile and a half below these ruins is a pretty high, but sloping slaty bluff; having come about nine miles, we arrived at a hunting camp of *Chactaws*, on the west side; who invited us on shore, treated us very kindly, and spared us some venison, bear's meat and oil. The afternoon being stormy, with hail and rain, we encamped at a small distance from them: the canes were not very plenty here, but the land rich: a great deal of the plant called *Indian Hemp* grows in this place; but the season deprived me of the satisfaction of knowing what genus it is of.

9th. At half an hour past eleven proceeded, and in an hour and a quarter we passed by *Chickianoeé*, a white bluff, with a savannah on its top, on the west side; it is upwards of seventy feet high above the water's level: we past several high bluffs, among which one is yellow like *Ochre*. We saw many bars also, and lagoons: having travelled twenty-four miles, came to camp at four o'clock P. M. on the east side, at the beginning of a steep slaty bluff.

10th. At half an hour past nine, A. M. proceeded; we went between high bluffs, and in two hours time came to the mouth of *Tuscaloosa* river from the east; and a little below it is the steep white chalky bluff, on whose top is a vast plain, and some remains of huts in it; the bluff is called the *Chicasaw Gallery*, because from here the savages used to annoy the French boats going up to the fort, or down from it. The river is hereabouts full of rapids, and bad passes: we came past a number of high bluffs, most of them chalky. At half an hour past three we past the mouth of *Sookhavatcha* from the west, and three miles below it came to camp, at four o'clock P. M. at the foot of the hill, where formerly the *Coosadas* were settled: this place is called *Suktaloosa* (i.e.) Black Bluff, from its being a kind of coal; it is a great thorough-fare for warring savages, therefore we took the usual precaution of large fires, and hanging our hats on stakes, which we had reason to think not in vain; for in the night we heard the report of small arms. This day we came thirty-six miles. It is worthy of remark, that although we have come near seventy miles from the ruins of *Tombechbé*, yet by land the distance is not above twenty-four, or twenty-five miles. The land here is very fine, and Mr. *Dow* told me, that he had lived here with the *Coosadas*, and that the common yield of corn was from sixty to eighty bushels per acre; that they increased horses and hogs to any degree they pleased, and that venison, turkies, and fish were uncommonly plenty.

11th. Last night and this morning, being rainy, we could not proceed till eleven o'clock. All this day we passed through the remains of the *Coosada* and *Occhoy* settlements, being all a fine tract of ground, of which much had been cleared; but it is now again overgrown with reeds; the grand, or publick plantation in particular is an excellent tract. At four o'clock P. M. we encamped on a little plain, under a bluff, where was a large hunting camp, to appearance about two years old: here we saw some stones, having been deeply marked by the savages, with some uncouth marks, but most of them being straight lines and crossed, I have since been led to conjecture, whether they were not occasioned by these people grinding their awls on these stones; yet they do not ill resemble inscriptions: this place is a pretty situation, and is near two miles below the deserted *Occhoy* town, which stood likewise near a black coaly bluff; the distance we made this day is twenty one miles, we had fine weather during the day; but the night was showery.

12th. At half an hour past eight A. M. proceeded. All this day we saw marks of great fertility of soil, and much tolerably high land: at half an hour past ten, we were at a creek called *Abeshai*, at a quarter past eleven at the last *Occhoy* field, by a creek called *Bashailawaw;* at eleven A. M. at the hills of *Nanna Falaya*, on the east side; which rise steep out of the water, about fifteen, or twenty feet, then slope up into very high short pine hills. Some parts of the rock are red, others grey. Here we were overtaken by very bad weather, from which took shelter; at half an hour past one, P. M. we encamped about three quarters of a mile below the hill, on the slope of a pretty high bank, where we found the remains of a camp, that had been occupied lately by white people; we came about eleven miles and an half this day; the rain continued till two o'clock A. M. next day, when the wind shifting to W. N. W., it grew excessively cold.

13th. At half an hour past ten A. M. proceeded; at one o'clock we came to a hill on the east side, with an old field on its summit; this hill is called *Batcha-Chooka;* here we found a notorious gang of thieves, belonging to the town of *Oka Loosa*, a town in the Chactaw nation; when we say their raft, we took them to be a Creek war party, therefore, being hailed by them, and not choosing to be shot at, we went near the shore; but on discovering who they were I refused to land; they still insisted we must, but my obstinate persisting to the contrary, disappointed their sanguine hopes of plunder; and after some altercation I proceeded. This day the marks of fertility of soil are not so uninterrupted as on the former days. Our weather was clear, and a strong northerly wind prevailed;—we came nineteen miles since morning, and encamped on the west side, in a low spot of ground.

14th. Last night the frost was severe; at a quarter past nine,

A. M. proceeded;—in half an hour's time we saw very high hills, at a mile, or better, from the river, seemingly covered with pine timber; these hills the savages call *Nanna Chahaws*. Here is a steep place above forty feet perpendicular out of the water, and another steep above it; the last is a grey slaty rock; this place is called *Teeakhaily Ekutapa*, and the people from *Chicasahay* had a settlement here before the war. About a mile and a quarter from hence is a remarkable white sand hill on the east side; four miles lower we came to *Yagna-hoolah* (i.e.) the Beloved Ground, which lies on the east side, and is very high, continuing above two miles along the river bank; its lower part is steep and of a whitish grey, and at the end above two hundred feet high, reckoning perpendicularly. A mile below this is a white sand hill on the west side; we saw the pine hills all this day, at various distances from the river, sometimes close to it, and the canes begin to diminish, and pine trees mix among the timber. The current for these two last days, is very considerably slackened, and the river widened to above five hundred feet. We came about thirty miles and an half this day, and encamped at four o'clock P. M. on the side of a kind of bluff, about six miles below a branch called *Isawaya*: all day cold, the latter part and night dull and hazy, at a quarter past nine proceeded. Eight miles below our camp we were at the mouth of *Senti Bogue* (i.e.) Snake Creek, having an island in its mouth, and coming from the westward. Two miles and a quarter lower is *Atchatickpé*, a large bay, or lagoon, on the same side: at this place is the beginning of our boundary with the Chactaws, running from the west till it strikes *Senti-Bogue*, and then follows the course of said creek, up to a certain sugarloaf hill, and so over to *Bogue Houma*, and *Bakkatané*. A mile below this, at the bending of the river is a bluff, but not very high, of a dark grey stone; above this it rises gradually sloping into a very high hill, variegated into small ridges. We saw many spots of pines, and some white sand hills; but in general the soil has a better appearance than yesterday. About an hour before we encamped, we came to the last rapids, or the first from below; here is a remarkable spot of yellow rocks in the western bank, beginning with a high, perpendicular, white rock, with some grass spots; it is above fifty feet above the present surface of the water; its top is level and shrubby; in the middle projects a remarkable lump, which, in coming down looks exactly like a buttress against a wall. At four o'clock P. M. we encamped on the east side in the low ground, above a mile below the rocks, having come thirty-one miles. The weather has been clear and cold all this day. Stout sloops and schooners may come up to this rapid; therefore I judge that here some considerable settlement will take place.

16th. Proceeded at half an hour past eight, A. M. Having come about fifteen miles, we saw the remains of the old *Wectumpkee* settlement: about seven miles below this, on the east side is an old rocky bluff, appearing to be sandy, and is covered with cedar trees. The river here is very crooked, and about six miles below on the west side, we saw a spacious old field, and a smoak in one edge of it; but nobody near it, and two miles lower down, hearing a rustling in the canes, we looked that way, and saw a savage in a war dress, lying flat; finding himself perceived, he got up and beckoned to us; but although we were within ten feet of him, we seemed not to have remarked him, upon which he lay down again: it is to be imagined that he was not alone, and that this was a war party, who had been at the smoak, in the old field, and having perceived us, had come to this place, knowing that here we must come near the bank; but seeing that we were no *Chactaws*, and thinking themselves undiscovered, they kept close. At a quarter past four o'clock, P. M. we came to a camp on the west-side, which we supposed by the boats, &c. to be occupied by white people; in which opinion we were soon confirmed. When they invited us on shore, we found they were one *Thomas Baskett*, with two white hunters, and some *Chactaws;* we were here well regaled with excellent meat, and very good bread, which being prepared in an excellent manner, was a noble feast to us. I purchased some bear, bacon and venison hams of them and staid all night at their camp; the distance we came this day was thirty-one miles.

17th. Embarked at half an hour past nine, A. M. and proceeded, accompanied by two canoes with savages: we soon past by some high pine hills on the east-side; and at their end, having come about two miles, we were at the little creek called *Apé Bogue oosè*, which is a spring so intensively salt, that the savages told us, three kettles of its water, yields one of salt. Having then proceded for four hours through low land on each side, we arrived at the place called by the French, *The Forks* being a lagoon divided into three branches, whereof the first is called *Apé Tonsa*, the second *Beelosa*, and the third *Caäntacalamoo:* here the savages left us; we still proceeded for half an hour more through low land, and then came to a large bay, at the end whereof begins the *Tomeehettee* bluff, where formerly a tribe of that nation resides; this is the first time we have the real pine barren butting on the river, it is very level. About five miles below this place, we came to the first islands that are of note; the land continues low and pretty rich: here we see the first summer canes. At six o'clock P. M. we came to the *Coosadas* bluff, having had the *Naniabé* (i.e.) Fish Killer's island above an hour on our west side; this place was the last settlement of the *Coosadas*, after they left *Sukta loosa;* and in little more than half an hour we were at the mouth of the great *Alibamo* river. We past the *Nita Abé*, or Bear Killer's Bluff on the left, and at nine o'clock P. M. we came to the north end of the island, which divides the branch called *Dog River*, from the west branch of the river. Here we staid all night at the plantation of the *Chevalier de Lucere*, but found only three or four old slaves and children, the whole of the able hands, and the overseer, being gone to make tar-kilns, so that we had but indifferent fare. We came this day forty-two miles and a half. N.B. Those islands are very fertile, and have a great many plantations on them, on the branches which lay out of our way, particularly on the *Taënsa*, and *Dog Rivers*.

18th. At a quarter past nine, A. M. proceeded, past several plantations, as well on the islands, as on the main, particularly *Campbell's*, *Stuart's*, *Ardry's*, and *McGillivray's:* at half an hour past eleven, A. M. arrived at Mr. Favre's house, where I staid in order to get some refreshment; this being the first Christian habitation I had been at since the 20th September, last year. Mr. *Favre* treated us in a most friendly, genteel and hospitable manner. At one o'clock some boats went up the river, which I heard were Mr. *Stuart's* people, with a provincial deputy surveyor, going up to ascertain the boundary between us and the *Chactaws*. At two o'clock, some gentlemen, among whom was Major *Dixon*, of the sixteenth regiment, and *Charles Stuart*, deputy superintendent of Indian affairs (to whom I described *Atchatikpè* and *Senti-Bogue*, where they were to begin) followed them: they proceeded up to Mr. *Stuart's* plantation, about three miles higher up the river. We had come seven miles and a quarter this day. In the afternoon it begain to rain, and all night was a prodigious storm of wind and rain, which I had the pleasure of weathering out under a good roof: here we found several families of *Chactaw* savages.

19th. At a quarter past nine, A. M. we proceeded, went past *Chastang's, Strother's,* and *Narbonne's* plantations, having chiefly pine land on the main, and the rich islands on our left all this day. Having gone five miles and three quarters, we passed by the ruins of *Fort Condè,* or old *Mobile,* and near six miles lower down we past by the ruins of a fine plantation, formerly belonging to the French Intendant at Mobile, now to Mr. *Lizars* at the same place: 4 miles and 3 quarters lower down we met with the first marsh, the river very full, we could not learn how far the salt water had its effect, the bay itself being fresh and good at this time; but Mr. *Dow,* who had been several times up and down this river, and had lived with the *Coosadas* for some years, assured me, that the tide was very visible at the old *Wetumpkee* settlements, and in extraordinary tides even as far as *Seekta Loosa,* where, during his residence on the spot, he has frequently seen it ebb and flow about an inch. We came this day thirty-five miles and a half, and at nine o'clock P. M. we arived at *Mobile.*

N. B. It is to be observed that the general course of this river is from north to south; but it is very crooked.

Journals are tedious, and I believe the above account will sufficiently given an idea of the country, else I might record more of this kind, especially a journey from *Pensacola* to *Manchac,* and down the *Mississippi,* from thence to the sea; but I will content myself with publishing a copy of a paper which was given me by Captain *Rufus Putnam,* with liberty to make what use I pleased of it. This gentleman was one of the committee appointed by the people of Connecticut, to explore the country about the *Natchez.* He sent a letter of the same contents to my very worthy friend Doctor J. *Lorimer,* at *Pensacola;* this paper will compleat the intention of my giving an idea of the country, and with it I will conclude this my first volume.

It is as follows:

"Sir!

"According to my promise, when I left *Pensacola,* I now send you the observations which I made of the *Mississippi* country, more especially that part we were sent to explore, and I choose rather to do it from extracts of the minutes I made in the woods, than by any general observation on them, and I am sure that your goodness will excuse every fault in my vulgar way of writing, since, if you have but the facts, you will be able to form a general account in such language as you please, and I shall begin my account first at the Natchez.

"26th April, 1773, arrived at fort *Rosalie*, which is built on a high eminence, that overlooks the whole country, about two hundred yards from the river; it was built of a heptagon figure, with one side fronting the river; each side, or angle, was about thirty yards clear, with two gates in the eastern angles. I took here the meridional altitude of the sun, by Davis's quadrant, as I did again on my return, and I make the latitude of this place to be 31° 15' north; from hence I also saw the range of high lands from *Loftus's* cliffs, and those on the east side of the *Hooma Chitta*, which is near twenty miles distant, the lands hereabouts are very uneven, no creeks, and much worn out.

"27th Visited *St. Catherine's* creek; found the lands very good, timber walnut, hickory, oak, ash, &c. Mr. *Thompson's* well was perfect good water, without stone or curve.

"30th. On the point above *Boyd's* creek travelled two or three miles into the country, found it mostly cane land, but subject to overflowing.

1st May. Arrived at Petit Goufre, where is a firm rock on the east side for near a mile, partaking (in my opinion) of the nature of lime stone, the land near the river is much broken and very high, perhaps three hundred if not five hundred feet high; about four miles farther up the river is a point of land somewhat low, but good, and stocked amazing thick with mulberry.

3d. About four miles up the bay one *Pierre* landed and spent some time in exploring the country, found the land to the N.N. W, and N. E. low, southerly the lands make higher; but broken into hills and vales, but then the low lands are not often overflowed,—white oak, live oak abound here, intermixed with copalm, and other timber, common on the Mississippi:—going up this to the fork, which is called seven leagues, we saw several stone quarries, in the bank some gravel bottom and sand banks, went unto a hill to view, found it broken as far as we could see, clay soil, and some gravel stones on the top of the ground.

"4th.. Arrived at the forks, where the river parts almost at right angles, and which is the biggest branch I cannot tell: travelled N. E.

about 3 miles and a half up, until we fell on a small creek that falls into the south fork of the *Basoune Pierre*, down which we came to our boat; this branch we found winding as the main creek; the lands we saw are clay and marl soil, not so uneven as we had before seen on the creek; arrived at the big black, *(Petite riviere des Teaux)* where part of our company set off by land to the *Yasoo*, and the rest of us by water, found many parts of the bank high enough to build on before we came to the *Yasoo* cliffs, which are very high. Nine miles up the river *Yasoo* found Captain *Enos*, and those that went with him, at a place where it is said the *French* formerly had a fort and settlement; the bank on the right hand, going up at this place, is a high ridge of rocks, with a fine spring of water, falling from the *quarne*, the *Yasoo* river is about four hundred, or five hundred feet broad, its course is east, about two miles and an half up N. E. to the above place, is a very dead, stagnated water, of bad colour, covered with scum, and abounds with aligators; its banks in some paces are high, but cut through in many places, by the overflowing of the *Mississippi;* about two miles and a half from the mouth is a pleasant creek, comes in on the right, to which we returned and encamped, from whence we made several tours into the country, found all this point between the *Yasoo* cliffs and the river, low and full of ponds.

"13th. Colonel *Putnam,* Mr. *Lyman* and myself set out about ten o'clock A. M. for the purpose of farther reconnoitring the lands between the old fields, on the *Yasoo* cliffs, we went up the *Yasoo*, near the old field, or French settlement, steering south; in about two miles, came to a dead creek, which we traced eastward, till we found it came out of the high lands; we then bore southward two or three miles, travelling by some cypress swamps, some lands not overflowed by the Mississippi, keeping the highlands in sight on our left; this brought us into an excellent flat piece of land, full of grass, and some cane, with oak, walnut, and other wood, common in the country, we discovered signs of water spreading over this land from the hills in several places, we steered east as we found the highlands, bore farther off, but was soon taken up with a mighty cane break; Colonel *Putnam* here climbed a tree, discovered highland at about a hundred rod distance, we travelled for it and arrived at it, in about two hours, with the utmost difficulty, in our way found a fine running spring of water; ascending the highest part of the hill I climbed a tree, from which I had a fine prospect, found the lands N. E. and S. to be hilly, but not mountainous, nor much broken, then returned to the spring aforementioned and encamped.

"14th. Arrived at the *Yasoo* cliffs, on the *Mississipi*, where we met the rest of our company, we came down under the high lands all the way, found the country good.

"15th. Mr. *Lyman*, and myself, went unto the cliffs, which shut quite down the river, on these hills I climbed two trees, and found the land make high north eastward, and S. S. eastward, bearing off from the river, but somewhat uneven, full of cane and rich soil, even on the very highest ridges, just below the cliffs, the bank is low, by which means the water on the *Mississipi* flows back and runs between the bank and high land (which range near north and S. S. E. to the *Loosa-chitta*, forming much low land, cypress swamp, and dead ponds, without one brook, or running stream, as Capt. *Enos* informed me, who went up that way by land.

"17. Went up the big Black *(Loosa Chitto)* which is at the mouth about eighty feet wide, but within, from eighty to a hundred and forty; in about a mile and a half came to high land on the right, but broken, a mile and a half further, the high lands make again on the right; spent much time in reconnoitring the country, saw several springs of water on the right, but none on the left; at eight miles past Mr. *Cluere*, an Indian trader; here the high lands come near the river on the left, and appear to be the same range that comes from the *Yasoo cliffs.*

"18th. Being fourteen miles up, lands make high near the river for two or three miles, but broken, though good and full of springs of water: in the afternoon made high level land on the left, (which is now reserved for the capital:) after reconnoitring the country round (which we found very good for a settlement, or building a town) we encamped.

"19th. Set out in the morning, up the river, reconnoitring both sides for four or five miles from the river, found the land good, not so much drowned, nor so uneven; after rowing the boat about six miles and a half further, we came to a rapid water, stone or gravel bottom, twenty or thirty rods, and in one place a firm rock, almost across the river, and much of it bare at this season (which is neither low nor high water) as confines the water to nearly twenty feet, and the channel about four feet deep.—I shall add no more by way of extract, and shall observe only, that I have said little about the *Mississipi*, as it is so generally known that I could afford you but little light in the matter; but have confined my extracts to the more unknown parts; one thing I would observe about the postage of the cross, above *point coupé*, which is, that I do not believe any such thing, as a communication across here, in high water. I spent two days up that creek, found it grow less and less, by branching out into small rivulets, coming from high lands, and not the least appearance of lakes, or drowned lands."

It may, perhaps, be agreeable to the inhabitants of East and West-Florida, who become purchasers of this work, to have a copy of such articles of the late treaty of peace, in 1762, as immediately concern them. I have therefore subjoined, by way of appendix, to this volume, the subsequent extract from the preliminary articles of peace between his *Britannick* Majesty, the *Most Christian* King, and the *Catholick* King; signed at *Fontainbleau*, the 3d day of Nov. 1762.

ARTICLE II. His *Most Christian* Majesty renounces all pretensions, which he has heretofore formed, or might have formed to *Nova Scotia*, or *Acadia*, in all its parts, and guaranties the whole of it, with all its dependencies, to the King of *Great-Britain:* moreover, his *Most Christian* Majesty cedes and guaranties to his said *Britannick* Majesty, in full right, *Canada*, with all its dependencies, as well as the island of *Cape Breton*, and all the other islands in the gulph and river of *St. Laurence*, without restriction, and without any liberty to depart from this cession and guaranty, under any pretence, or to trouble *Great-Britain* in the possessions above-mentioned. His *Britannick* Majesty, on his side, agrees to grant to the inhabitants of *Canada*, the liberty of the *Catholick* religion, he will in consequence give the most exact, and the most effectual orders, that his new *Roman Catholick* subjects may profess the worship of their religion, according to the rites of the Roman Church, as far as the laws of *Great-Britain* permit. His *Britannick* Majesty further agrees, that the French inhabitants, or others, who would have been subjects to the *Most Christian* King in *Canada*, may retire in all safety and freedom, where ever they please, and may sell their estates, provided it be to his *Britannick* Majesty's subjects, and transport their effects, as well as their persons, without being restrained in their emigration, under any pretence whatsover, except debts or criminal prosecutions; the term limited for this emigration being fixed to the space of eighteen months, to be computed from the day of the ratification of the definitive treaty.

ART. VI. In order to re-establish peace on the most solid and lasting foundations, and to remove forever every subject of dispute, with regard to the limits of the *British* and *French* territories on the continent of *America;* it is agreed, that for the future the confines between the dominions of his *Britannick* Majesty, and those of his *Most Christian* Majesty, in that part of the world, shall be irrevocably fixed, by a line drawn along the middle of the river *Mississipi* from the source, as far as the river *Iberville,* and from thence by a line drawn along the middle of this river, and of the lakes *Maurepas* and *Pontchartrain* to the sea, and to this purpose the *Most Christian* King cedes in full right, and guaranties to his *Britannick* Majesty, the river and port of *Mobile,* and every thing that he possesses, or ought to have possessed, on the left side of the river *Mississipi,* except the town of *New Orleans,* and the island in which it is situated, which shall remain to *France;* provided, that the navigation of the river *Mississippi* shall be equally free, as well to the subjects of *Great-Britain* as to those of *France,* in its whole length and breadth, from its source to the sea, and that part expressly, which is between the said island of *New-Orleans,* and the right bank of that river, as well as the passage both in and out of its mouth: it is further stipulated, that the vessels belonging to the subjects of either nation, shall not be stopped, visited, or subject to the payment of any duty whatsoever. The stipulations in favour of the inhabitants of *Canada,* inserted in the second article, shall also take place, with regard to the inhabitants of the country, ceded by this article.

ART. XIX. His *Catholick* majesty cedes and guaranties in full right to his *Britannick* Majesty, all that *Spain* possesses on the continent of *North-America,* to the east, or to the south-east of the river Mississipi, and his *Britannic* Majesty agrees to grant to the inhabitants of this country above ceded, the liberty of the *Catholic* religion: He will in consequence give the most exact, and the most effectual orders, that his new *Roman Catholick* subjects may profess the worship of their religion, according to the rites of the *Roman* church, as far as the laws of *Great-Britain* permit. His *Britannick* Majesty further agrees, that the *Spanish* inhabitants, or others, who would have been subjects to the *Catholick* King, in the said countries, may retire in all safety and freedom, &c.

FINIS.

*A LIST OF SUBSCRIBERS, whose Names came
too late to be prefixed to the Work.*

(B)   Mr. Elias Beers, New-Haven, Book only.
Capt. Nathan Briggs, Newport, R. Island.

(C)   Mr. Christopher Champlin, Newport, R. Isl.
Mr. George Champlin, Newport, R. Isl.
Honrorable John Collins, Esq; Newport, R. Isl.

(D)   Benjamin Douglas, Esq; New-Haven.

(E)   Capt. Robert Elliot, Newport, Rhode Island.

(F)   Mr. Samuel Fowler, Newport, Rhode Island.
Capt. Laurence Furlong, Newbury-Post.

(G)   Capt. Caleb Gardner, Newport, Rhode Island.
Capt. George Gibbs, Ditto.
Capt. Robert Grant, New-Haven.

(H)   James A. Hillhouse, Esq; New-Haven.
Henry Hunter, Esq; Newport, Rhode Island.

(M)   Francis Malbone, Esq; Newport, R. Island.
Col. John Malbone, Ditto.
Henry Marchant, Esq; Ditto.
John Mawdsley, Esq; Ditto.
Capt. Charles Moore, Ditto.
Capt. William Morony, Ditto.
Thomas Mumford, Esq; New-London.

(P)   Samuel Holden Parsons, Esq; New-London.
Mr. Henry Pelham, Boston.

(R)   Mess'rs Reak and Okey, Newport, R. Island, (25 Copies of the Book.
Abraham Redwood, Esq; Newport, R. Island.
Capt. James Rice, New-Haven.
Mr. James Rodman, Newport, Rhode Island.

(S)   Capt. Thomas G. Steele, Newport, R. Island.
Mr. Robert Stoddard, Ditto.
Mr. Thomas Stewart, New-York.

(T)   John Tillinghast, Esq; Newport, R. Island.
Mr. Nicholas P. Tillinghast, Ditto.
Mr. William Tweedy, Ditto.

APPENDIX

APPENDIX

The following directions to navigators were originally intended as an appendix to the whole work; but I have since found it requisite to publish them sooner; and therefore end this first volume with that necessary part of my work.

These directions I beg leave to affirm, are entirely founded on my own experience; the first spur to this undertaking was the loss of a vessel in 1766, on the coast of *Florida:* by which I received a wound in my circumstances, which is as yet far from being healed. I had before this, in the same vessel, at a time when I was myself on board, struck on the south part of the *Dry Tortugus,* when having accidentally a tolerable pilot with me, I passed for the first time, inside of the reef; and made my first observations towards this present work: Since that I have pursued it partly during the opportunities I had at the time when I was a provincial deputy surveyor, and partly when I was appointed deputy for the southern district of *America;* and had the honor of commanding the vessels on that service. Likewise, while employed by the superintendant in West-Florida; but far most at my own great expence, and fatigue. For even when I was employed, I can safely affirm, that when I could do it without detriment to the service, and at times when I did that business, in a manner which is vulgarly called *by the jobb,* I have done more than was desired of me; which, however, has never been the occasion of any favor being bestowed on me, any more than *you did well, and I am glad you did it.* Therefore I am under no manner of obligation to any of the LITTLE GREAT ONES, who have occasionally used me, (sometimes as the monkey did the cat) and this production being undoubtedly my own, as such I have a right to publish it.

Thus having shewn how I got my information, I declare myself free of copying or compiling from prior works, except in that part lying west from the *Missisippi,* which is taken from *French manuscript draughts;* and the shape of the coast and bays eastward from *Pensacola,* to *Cape St. Blas,* which I have followed from draughts occasionally seen in the hands of Mr. *Stuart,* (the super-intendant) and my good friend Dr. *J. Lorimer*, which draughts I take to be the work of Mr. *Gauld*, a very able and accurate observer; but though I kept the shape, the distances are different from his; and are such as I have reason to believe nearer the mark, during a long and tedious cruize I made on this coast in 1771. As for the soundings, I never had his draughts long enough, to have an opportunity of perusing them, but they are such as I have obtained during the same tedious cruize; as for those from *Cape Blas*, round to *St. Mary's*, I have obtained them during the regular pursuit of gaining surveys of the coast for above seven years.

And as I have in the opinion of many of my friends in my proposals published before I begun this work, reflected a little too severely on the wretched map-makers who have attempted to explain this so intricate navigation, it is still my opinion, that such exhibitions can not be too much despised, when attempted on so loose a foundation, as has been generally done; and we must still form a worse idea of those who impose on the public against their better knowledge. Both the one and the other has been done. I therefore publish the present work with no other view than that of pointing out the sad mistakes of the scandalous maps daily published by persons who either not at all , or at best scarce ever saw the ground they treat of, and thus merely for the sake of catching a penny, endanger such numbers of lives, and such a vast deal of property;—such imposters are the shame of the nation to which they belong.

I have carefully avoided the change of well-known names of places; but preserved the old ones, except only in two or three places where the name was not well known, or whether there was none at all. Nothing can be more absurd, or productive of confusion, than the assuming new and fantastical names in places of so much danger; yet the author of a certain pamphlet, published two years ago, has done this at no small rate.

And for the safety of navigation, I think it a duty incumbent on me, to declare that the soundings of what the said author calls the *Newfound passage* or *channel* (altho the knowledge thereof is as old as that of *America* itself) are marked much deeper than they really are; and whoever depends on that pamphlet, will find himself much imposed upon; nor do I think my work to be absolutely clear of faults; no, but I venture to say, that the errors are very few; and as they are unknown to me, therefore, whatever mistakes

may be discovered, if communicated by men of knowledge, I shall
gladly receive, and willingly amend them.

Vessels bound from *Europe* to *St. Augustine, Georgia,* or *South-
Carolina,* would (in my opinion) shorten their passages considerably,
by making the south end of *Abaco,* or the *Hole in the Rock,* in lat.
26° 4' and from that run W b S to make the *Berry-Islands;* from
thence W b N, or WNW, till they get into the gulph stream; the only
precaution to be observed is to steer to the westward of north, after
you are clear of the *Grand Bahama* island, because the bank stretches
N b W nearly, and the current sets partly on the NW part of the
bank; particularly near the *member* or *memory rock.* You may, in
case of necessity, venture on the bank, to the south of the rock, and
beat up to the eastward, till you bring the rock to bear W N W, and
you will not find less than five fathoms water:—The tide ebbs and
flows here about five feet, but you cannot come out on the north of
the rock; however, by keeping in the stream, I dare venture to say
you will find a visible benefit in coming this way.

DESCRIPTION OF THE STATION FOR CRUIZERS WITHIN THE FLORIDA REEF

The first of these is at *Ceyo Biscayo,* in lat. 25° 35N. Here we
enter within the reef, from the northward; directions for an easy
approach to the entrance will be given hereafter: Suffice it now
to speak of the entrance itself;—for about five leagues north of the
key, the ground is very foul, and looks frightful, but there is no where
less than 3 fathoms, though by keeping out about 5 or 6 miles from the
shore, you will find generally 5 or 6 fathoms, on a fine sandy bottom;
and when you approach the reef, you may haul in, observing to leave
the reef a large piece without you, for it has many bad sand bars just
on its inner edge: you will not find less than 3 fathoms, any where
within, till you come abreast the south end of the key, where there is
a small bank of 11 feet only, give the key a good birth, for there is a
large flat stretching off from it. At the south end of the key, very
good water is obtainable by digging, but at a time when by accident
of drought or otherways, the wells yield none, then the watering
places either in the grand marsh, about 10 miles to the W by S; or
the river *Ratones,* about as far to the NW of the key, may be always
depended upon; at the watering place on the key, is also an excellent
place for careening of vessels, not exceeding ten feet draught. All

these advantageous circumstances, joined to some of the same kind at *Bemini* islands, make this an excellent station for cruizers to watch every vessel bound northward; on the bank without the *Bemini* harbour is good anchorage for large vessels. The method I would recommend, is to have two vessels, one to serve as a tender to the larger one, and while the one is on the *Bemini* side, the other ought to be on the *Florida* side; by this means (the passage being only 15 leagues wide) no vessel could come through during day, without being seen by one or the other; and as it is always possible to get out-side of the reef, except with a strong N, or N E wind, and flood tide; I need not direct seamen how to proceed, in order to speak with any vessel they chance to see. The many breaks that are in the reef, make it not improbable that a passage might be found through even with a N or N. E. gale. As for the safety in laying within side, every seaman knows that a reef makes a safe harbour; good ground-tackling is the principal point; should a chace lead you far north in the stream, there is no difficulty in reaching your station again; all that's necessary, is to make the land as soon as you can, only avoid *Cape Canaveral*, in lat. 28°: 30'; to the south of it, you may make pretty free with the shore, except in the bight north of *Rio d'ais*, where there is but 12 feet a league off. To the south of 26°: 50', to about 26 there is 7 fathoms at the very beach: here you must hug the shore close, because the stream almost touches the beach: I was once in great danger hereabouts by the ill working of the vessel I was in, and had just time to let go an anchor in 9 fathoms, and when the vessel rode to it, it would have been an easy matter for a man to have jumped off from the tafferel on the beach; the beach here is for the most part smooth enough for a canoe to land; there being little or no surf on it in good weather.

In the year 1773, I came a passage from *Mississippi*, on board the schooner *Liberty*, commanded by capt. *John Hunt*, we had the misfortune to be over-set at sea, and I conducted the wreck into this place, when having lost our boats and caboose, with every other thing from off the deck, we nailed together three half hogsheads, in which a man and a boy went on shore, and brought us sand off, to make a kind of caboose. I mention this the more, because there is water obtainable in almost every place on this beach, in case of necessity. But to return to my subject, to keep the shore on board is the way to get to the southward here; and as on the soundings, the current is chiefly ruled by the wind, it is preferable to come to an anchor with a southerly wind, instead of standing out; for the stream will run you out in such a manner, that you will find it difficult to regain in four or five days, the ground you lose in one. Yet I would avoid anchoring as much as possible, because the ground to the south of 27° is almost every where very foul, particularly, if you cast deeper than 15 fathoms: by observing this method, you will find it an easy matter to regain your station to the southward.

At this place there is vast abundance and variety of fish, both in the creeks, and out-side at sea, particularly *groopers* are in great plenty, *king-fish, Spanish mackrel* and *Barrows* are also often caught towing; and if you have one or two good hunters on board, you may always be provided with plenty of venison, turkies, and bear meat, which are all excellent refreshment. There are for the most part deer on the key, and sometimes bear; but by sending to the main, you may depend on finding every kind; in winter, duck and teal abound in the creeks; turtle is very plenty; in short, I know not any place better calculated for a cruizer, then the spot I have now described; add to this, that no fish is poisonous on the *Florida* shore, not even the *amber-fish;* but on the *Bahama* side, precaution is necessary; and the loggerhead turtle is never rank of taste here. The watering place on *Bemini*, is near the east end.

The second place is at the south end of *Key-Largo*, near *Sound-point, Tabona*, and *Rodriguez;* in lat. 25°: 3'.N. Here is the eastermost pitch of the reef, as *Sound-point*, is of the land, and therefore more proper to be called *Cape Florida*, than any other part of the shore, particularly as Sound-point is a part of the main land. *Key-Tabona* is a small low key, lying about one mile and a quarter without the main keys; and two miles and one quarter north of this, lies another something larger, called *Key Rodriguez;* these two keys are a good mark to know the entrance by, they being both overgrown with mangroves, appear of a darker color than the land before which they lie. When the southermost of *Key Tabona* bears W. 3½ leagues off, then you are near the south part of the outermost reef, which is in general almost dry, and in a few spots entirely dry; run within musket shot to the southward of this dry reef, and then steer east, as you go in, and you will have 5 fathoms and upwards; run in on your lead till you begin to see the inner heads, and then draw northerly till you shoot within the outer reef; where you may lay pretty safe; however, I would recommend to have a pilot for the first trial of this entrance, among the *Providence* people such may be had; yet it looks more frightful than it is dangerous; at this place, nine vessels out of ten, that come through the gulph heave in sight, by reason that this is the land people generally strive to make, coming from *Cuba;* on this account also, nine out of ten, that are cast away, are lost here; if you have a tender, she may sometimes steer over to the bank, about the riding rocks, or *Orange-Key*, yet I think it can be but of little use, except when the season of westerly winds is. The channel from bank to bank, is here about 14 leagues a-cross; but to reach the bank, you ought to steer S. E. over from this station.

If from this place to chace should lead you northward, your endeavours ought to be to make *Isaac's Rocks*, or from *Key-Biscayno*, above described, to fall in with the *Bemini*, which is by no means difficult; the last is the most eligible, because the bank is broad, whereas at *Issac's Rocks* you have no soundings in less than a cables length from the shore, and consequently if you cannot make good a south course, you will be obliged to anchor on the north side of the rocks, and thus delay much time; but in making the *Bemini*, which lies nearly south 7 leagues from the rock, you may keep on soundings on the bank; the sounding however, as far as *Beak's-Key*, is not very wide. *Beak's-Key* lies S½ E7½ leagues from *Bemini;* from *Beak's-Key*, to the *Riding Rocks*, and *Roques*, there is working room plenty, and good anchorage; at *Orange-Key* is a good road to the S and S W of it; this key lies S S E 8 leagues from *Beak's-Key;* from *Orange-Key* it is an easy matter to get the *Double-Headed-Shot;* the way is to sail early in the morning, steering S. W.; when you have run about 10 leagues on that course, you may begin to look out for *Key-Sal* bank; where, in case of need, you may anchor any where; the *Dog-Keys*, and *Dead-Man's Keys*, shew themselves easily, lying on the N W and W part of this bank; they are bold too, yet dangerous; the soundings are narrow near them from here you may either make the *Cuba* shore, or run a-cross W N W, about 14 or 15 leagues, which will bring you in with the Florida reef, to the southward of your former station.

You may from *Key-Biscayo*, come southwards along the *Florida* shore, on the out-side of the reef; but as the sounding is very narrow, this way is by no means adviseable; except in fine and moderate weather, when you may keep your boats out, for the winds hereabouts being chiefly in the eastern quarter make the reef a very dangerous lee; whereas the *Bahama* side is perfectly safe. From *Key-Tabona*, to the N W about a mile lies a very small key called *Key-Palombo* or *Dove-Key;* which in wet seasons affords good water, but that is seldom; at about ten miles S W from *Key-Tobona* is *Matacombe*, here are several never-failing wells of excellent water in a rock; whether these are natural, or artificial is hard to tell: should these wells by accident of extreme drought happen to fail, which I believe next to impossible, you would in that case be obliged to send for water to *Key-Biscayo;* which is near 20 leagues from here. Fish and turtle are in as great plenty here, as at *Biscayo*. *Matacombe* yields a few deer of a small kind; but large deer, bear, and turkeys are not to be had without going to the main for them. I was once in great want of provisions at *Matacombe*, and sent a hunter with a boy in a skiff to the westward, at *Sandy-point* or *Cape- Sable;* whence he returned in a few days, with thirteen large and very fat deer, properly salted and cured, which were excellent provisions for us for several days.

The third station is at *Cayo-Huesos*, commonly called Key-West; this place is best calculated to watch those vessels which come from *La Vera Cruz, Campeche, Sisal,* &c. Few of those that are bound through the gulph come this way. There is a harbour, or rather a road stead inside of the westermost point of the key; but it is very unsafe in northern and western gales; and the ground is foul; so that in case of northerly or northwesterly winds, it is best to lay outside; therefore at the season when *Norths* are frequent, it is not advisable to lay within side, except for the purpose of hiding yourself: there is 20 to 24 feet going in to this place; and the way in, is to keep close on board the west side of the sandy key, which lies about 9 miles to the S E of the west point of the key; draw close enough alongside of it to chuck a biscuit ashore; and then steer directly for the west point of the great key, which point is bare, except one bush of trees which you will see on it: when you come close in, give it a birth, on account of a small bank lying on the out-side of it; to go further in you will see a black spot, being a small reef to the north of you, avoid this and shoot in to the eastward; your eye will be your guide; on the point are wells, but the water is indifferent; half a mile to the eastward is a path leading to very good water; on this key we find numbers of deer (of the small kind) generally very good: this place lies right in the way of the *Spanish* ships, from the westward; on account that these ships in coming from the western part of the *Mexican gulph*, chuse to run out of the way of the eastern or trade wind, and for that reason must go as far north as the coast of *West-Florida* will allow them; thus they are forced on an attempt to make the *Tortugas;* which often leads them further east, and then the currents generally render it difficult for them to get out, unless they know *Boca Grande* passage: for this reason a vessel in the proper season, cruising to the north of this key, and those that lay to the westward of it, can hardly fail of meeting them, and the ground is as safe as can be wished for; there being as regular soundings just north of those keys as are to be found in any part of the world; one may perfectly trust to his lead: The *Spanish* fishermen on the *Florida* shore all touch here in their way home: This key lies in lat. 24°: 29' N: long. 81°: 15' W: from *London*. In treating of the navigation within the reef, I shall repeat this account somewhat more at large. But will now proceed to give an account of the

COURSES ALONG THE EASTERN SHORE OF EAST-FLORIDA

From the mouth of *St. Mary's* to the mouth of *Nassau* river lies the island *Amelia*, which is a low even coast with sand-hills, and to be known by a detached hummock of trees on the south end; the shore is pretty bold too, except at the two ends, where the bars of said two rivers trench off a great way; the course is S b E, and N b W, the distance about 4 leagues; the bottom is a sand ground, the harbours on both ends are spacious, but St. *Mary's* is the safest; there is also a passage within, fit for small craft only; the bar at St. *Mary's* has 9 feet of water at low spring tides, and in going in the south shore is the safest and best to be trusted; the bar at *Nassau* has generally about 8 foot water on it, but is subject to shifting; the tides rises about 7 feet, and an E S E moon makes high water here as well as in most places along this coast; a plan of these two inlets taken exactly from my original survey of this island, was sometime ago published by Capt *Fuller*, who sounded the bars; but he has placed the soundings rather too deep, both within and without.

From the mouth of *Nassau*, or south end of *Amelia*, we proceed S S E, along the shore of *Talbot* island, which forms a bay; the distance is nearly a league and a half to a narrow inlet, between *Talbot* and *Fort-George* islands; this is scarce fit for boats:—Continuing our course S S E along *Fort-George* island one league more, we arrive at St. *John's* river's mouth; there are likewise two passages from *Nassau* in land, going by *Sawpit-Bluff,* and *Cedar-point* into St. *John's;* but will not at common tides allow of any thing more than 4 foot draught of water to pass; one of these called the *Two Sisters,* the other *Hannah Mills's Narrows;* some spots in these run quite dry; this mouth of St. *John's* river is easily known by a remarkable sandhill, called the *General's Mount,* a little to the south of it; there is at this inlet a dangerous bar, never at the highest tide (in moderate weather) admitting of any vessel drawing more than 9 feet; this bar is very subject to shift, so that directions for entering here, might tend to endanger the vessel, whose master should attempt to follow them; the best direction therefore, which I shall attempt to give, is to send a boat in to sound the bar, unless a pilot chance to come off; from this entrance, to about one hundred miles up this river, (of which 30 go to the westward, and the rest to the south); you may go in any vessel which the bar will suffer to enter.

Still going S S E from here, for 10½ leagues (equal to 36 statute miles) the distance measured on the land, we meet with the bar of St. *Augustine*: this bar is of the same kind as that at St. *John's;* therefore my best advice is to wait for a pilot, and trust to him; when the southern channel is open (such as it is laid down in the map) it is deepest, safest and easiest of access; there being near 18 inches difference in depth, between this channel and the northern one, when that is open; the beach between St. *John's* and St. *Augustine,* is even and strait, except a hill about 5 leagues S S E from St. *John's* (where we find three springs of fine fresh water), which is a little higher than the rest of the sand hills; this place is called the *Horse-Guards,* from General *Oglethorp's* posting a command of horse here, during his expedition against this country. This is the first place from *Long-Bay,* in *South-Carolina* southwards, where the inland navigation is interrupted; we may however, by going up St. *Pablo's* creek, in St. *John's* arrive near the head of the *North-River,* I think within four miles; and a small boat may with little difficulty be hauled over from one to the other. St. *Augustine* bar, is readily known from sea by the signal house on St. *Anastasia* island. This Beach of St. *John's* is tolerably bold, the soundings are regular; and the bottom is generally a fine white sand: but when you approach the south end, be sure of giving it a good birth, on account of the bar, which tends a long piece, (at least one league) north, and east of.

Time of high water at full and change here is half past 7 A. M. or in other words, an E S E, or W N W moon makes high water.

From the bar of St. *Augustine*, to the bar of *Matanca* is near 6 leagues the course S E b S, the soundings, beach and bottom similar to the above. *Matanca* bar will allow a vessel of 7 foot draught to come over: but as there is an inside passage from St. *Augustine*, for 5 foot draught, all the business is done that way, and scarce ever a vessel is bound in here. *Matanca* bar is known from sea by the fort, which shows white in a clear day, when the inlet bears W 3 leagues off.

Not much above half a league S E b S from *Matanca* is *el Penon*, a small inlet, scarce fit for boats to enter.

Continuing the same course for somewhat more than 16 leagues (the measured distance is nearly 57 statute miles on the land) we fall in with the entrance to *New-Smyrna*, or *Muskeeto Lagoon;* the map shews how far an inland passage is here practicable; the beach is much the same in appearance, the coast middling bold too, the soundings regular, and the bottom sand; with now and then some shells, and sometimes green mud.

The following are the position of the above places on the globe, according to my observations, mensuration and calculations.

	lat.	lon.	Wfr.	London.
St. *Mary* Inlet.	30° :	43' :	82°	
Nassau Inlet,	30° :	30' :	81° :	57'
St. *John's* Inlet,	30° :	20' :	81° :	53'
St. *Augustine* Bar,	29° :	52' :	81° :	35'
Matanca Bar,	29° :	37' :	81° :	28'.
Muskeeto Bar,	28° :	55' :	80° :	59'.

Next we meet an even smooth beach, from here to *Caneveral*, excepting *Mount Tucker*, which lies about 5 leagues south of the entrance of *Muskeete* River; this beech extends for 52 statute mile (equal to 15 English leagues) on a course S E b S: and here we find regular soundings as above, until we come to the cape which lies within a large and dangerous shoal, leaving a small channel between it and the land, as the draught expresses; the outer pitch of this head land lies in latitude 28° 16, and longitude 80° 28' W. from

London; and the pitch of the shoal extends 8 leagues east from it: in so much that when you are on the outer breaker, land is scarce visible from it; this remarkable shoal was never heretofore laid down in any map or chart for marine use, yet I am sensible that it is very dangerous; and I believe many vessels have been lost on it, the circumstances of whose misfortunes were never heard of; the wind on blowing a long and violent gale from E N E to S E makes this shoal almost unavoidable, to a vessel sailing in this latitude, not having a very large offing.

From this cape the shore stretches nearly south, to opposite the mouth of *Saint Sebastian* river in lat. 27° 56, then S E b S 8½ leagues to the *Tortolas,* (some hilly knowls so called) in lat. 27° 33; off this are some heads of rocks under water; 8½ leagues further (7 of which continue in a S E b S course, the other 1½ about S b E ½ E) we find the entrance of *Aisa Hatcha* or *Indian River,* from the shoal of *Caneveral* to within 2 leagues north of this river the coast is flat and treacherous; it is remarkable on account of the immense number of palm trees whence it has acquired the name of *Palmar de Ais,* or the palm grove of Ais: this inlet is the mouth of a lagoon lying north of it, communicating with a sound to the southward; this entrance lies in lat. 27° : 15, long. 79° : 56, and has a very shifting bar, sometimes not admitting a boat, and at other times 6, 8 or even 10 feet water has been found on it: therefore whoever has business here must send a boat to visit the bar, before he runs in, when in, it is a safe harbour; but a vessel must be moor'd, the tide ebbs and flows 5 foot, and runs with violence, the Spaniards come here for the purpose of fishing, and the quantity of fish and oysters found here is amazing; off this bar the sand has a peculiar quality of rubbing cables to pieces in the bent, insomuch that I never lay 24 hours off this place at an anchor, without wholly or nearly losing an anchor; though the weather was fine, nay I have often in less time found one or two strands quite chaffed off, I could never find what this was owing to, the sand is a very fine quicksand, nor do I remember ever to have found the ground foul, except close in, where there are some stones, and the remains of a wreck; in this lagoon and sound many streams of fine fresh water empty themselves; the principal are the rivers St. *Sebastian* and St. *Lucia* as the map shews.

From the entrance of *Indian* river the coast stretches S 26 E or nearly S E b S ½ S 11 1/4 leagues, (answering to nearly 39 statute miles as measured on the beach) it makes an island, at whose South end is an inlet into the above sound, the Spaniards call it *Hobe*; it will only allow of five foot draught, this inlet lies in lat. 26° 46' long. 79° 40' west from London: to the north of it, on the point at its entrance, lies a remarkable spot of rocks on the beach, as there is also at 1½ leagues to the northward of said entrance, and 3 leagues further north are several high blue or black rocks standing on the beach, these rocks make this part of the coast remarkable, as does likewise a hill inland full of white spots, a little to the north of the rocks inland; this hill is called the *Bleech-Yard*. A small reef just under water about half a mile from the shore abreast of the high rocks, forms here in a convenient little harbour for boats, there are two wells of excellent water in a little meadow, back of the sand hills, near a mile to the south of the high rocks; the river St. *Lucia* likewise shews its mouth over the sound, when you stand on the northermost high rock: the wells have casks in them, &c this is in my opinion a very good watering place for vessels who having had a tedious passage through the gulph, are in want of a supply of this necessary article; this island affords abundance of turtle, venison, and bear; as likewise numerous quantities of wild ducks, each in their season, besides *cocoplumbs* and *palm cabbage;* and on the main opposite, turkeys are abundant. The northermost of the high rock lies in lat. 26° 56. N. the coast is even, bold too; and the soundings regular.

We have now reached the southern extremity of the great bank of regular soundings which lies before the coast of America, and here the gulph stream comes very near the beach; the color of the water changes from a muddy green, to a beautiful saxon blue.

From *Hobé* inlet we find the coast trenching about S 20 E or nearly S S E for about 3½ leagues to a high ledge of rocks, out of which a large stream of fresh water rushes into the sea, a little to the north hereof is a small reef with about 2 fathoms water on it, where vast quantities of groopers, snappers, amber-fish, porgys, margate-fish, rock-fish, yellow-tails, Jew-fish, &c. may be taken; from this reef you see a high mount of sand & rocks a little to the north of you, three miles to the north of which is good fresh water, at the back of an high sand hill; as there is also a little more than half a mile to the southward or said rocky mount, this mount we called Grooper-Hill, it is a remarkable land, *cabbage* trees, *cocoplumbs*, and sea grapes are here in abundance, as are venison and other meat;

so that in fair weather a vessel may refresh here agreeably; the soundings along all this coast are plainly and truly marked in the map: the coast from here runs very nearly south, & the beach is bold too, to within 2 leagues of *Rio Del Medio*, it does not vary above half a point either way from this general south course all the way to *Cayo Biscayno* in lat. 25° 35. lon. 79° 36 W from London; as it is necessary for vessels bound southward to keep this beach close on board, I shall describe it particularly, in hopes that this treatise may prove a certain guide in every part of this difficult navigation.

Five leagues to the south of the above point or ledge of rocks, out of which a large spring of water issues; there is another ledge of rocks, and 1½ mile to the north of it stands a very tall tree by some fresh water, a large quarter of a mile to the south of this tree is a low sand hill, full of dry trees, whereby it may be known.

This last ledge of rocks is 1½ miles in length, somewhat more than 2 miles south of these, is another point of rocks, these are very low; 1½ miles further is a high cabbage grove, and about 3½ miles further south is a point of high rocks near 1/4 of a mile long, 2 3/4 miles further yet is another high point of rocks about 1/4 of a mile long, & half a mile south of this is the mouth of *Rio Seco* which is very seldom open, having a narrow bar of dry sand before it, the coast forms here a small kind of cove, which makes a good road stead for small craft, this river is in lat. 26° 16N long. 79° 35; 5 miles to the south of Rio Seco is another point of rocks, and south of it a small bite, and half a mile further is the mouth of *Rio Nuevo*, which is about 1/4 of a mile wide and generally open but shallow, here Jewfish are very abundant both within and without the river; 2 1/4 miles south of this river's mouth are five tall cabbage trees on the pine land; and 4 1/3 miles further is a thick scrubby point, and the sand hills in general are high and covered with thick bushes; here the shore is no longer so bold as before, but a bank of soundings about 5 leagues broad begins to cover the beach, and this (in my opinion) is the true beginning of the great Florida reef, 5 miles S of this is the north end of a marshy point which extends 1/4 of a mile to the mouth of *Rio Del Medio*, which is likewise a quarter of a mile across and shallow but very full of Jew-fish. From this river the beach has nothing observable on it except that its hills are higher than further north, and are covered with shrubs and trees. At 10 miles south of the river, is a remarkable spot of palm-trees. We see water over the land, and the coast is flat, there being not above 12 foot within a mile and more of the coast & the bottom solid rocks, about 8 miles south of the aforesaid palm trees is the mouth of *Boca Ratones* which will admit small craft, but has a reef stretching a considerable way to sea.

South of *Boca Ratones* about 5 miles, is the south point of an island, which has sillily been called *Cape Florida,* but since has acquired the name of *Fools Cape,* south of this point is a channel of a mile in width, having 6 foot water at the best of times, the opposite land on the south of said channel is the north point of Key Biscay, which is near 5 miles long from north to south; this key is particularly described in the beginning of this tract, as is also the entrance into its harbour, and its advantagious situation for cruizers, likewise the watering places on and near it; directly east from it 6 miles lie the *Fowey* rocks, which are the first dry spot on the reef, and have many bad bars within them, the key has likewise a bank lying off of it, but by good management in giving the key about 1½ or 2 miles birth you are sure to keep in a good channel, and till you come abreast the *Soldier Keys* or *La Parida* you will not have less than 16 foot water.

La Parida y su Figuelo, or as we call them *Soldiers Keys* are two low spots of mangrove on a bank, their distance from the south end of Key Biscay is 3 miles and they are inaccessible to any thing but a boat, the map will clearly shew their situation. Let it be observed that besides the watering places above mentioned, there are many more and that the beach will almost in every part yield good drinkable water for digging, provided the sand does not cover clay, wherever clay appears on the beach the labour would be needless.

Having now particularly and minutely described the shore, and the manner of going close along it, in order that this little tract may be as much as possible a perfect coasting pilot in this part of the world. I shall subjoin some

DIRECTIONS TO SAIL FROM ST. AUGUSTINE DIRECTLY TO THE ENTRANCE INTO THE INSIDE OF FLORIDA OR MARTYR REEF, AT FOOL'S CAPE.

As soon as you are clear of the bar of St. *Augustine,* steer SE easterly, the distance of 8 leagues on this course will bring you within about 4 leagues right abreast the *Matancas*, in 10 fathoms water, upon a bottom of fine dark grey sand. In continuing from here S E b S you will go along shore at an offing of 4 or 4½ leagues; in your way you will meet with soundings from 10 to 15 fathoms, chiefly on various kinds of sand, as the charts fully shew; and sometimes you will meet with a very soft greenish mud, in which the lead will sink to over the strap. All the soundings are in the charts laid down in feet, my lines of which I made use, being all marked with that measure.

The above offing you may with safety keep in with vessels of any draught for 25 leagues to the S E ward of the *Matanca*; you will begin to get shells among your soundings at 20 leagues, chiefly white and black, sometimes mixed with black and grey sand, but mostly the shells by themselves, very seldom sand alone, if you chance to find sand only, it will be a coarse grey kind; at 25 leagues as above, you will begin to get red shells, this is a true sign that you approach the Cape *Canaveral* shoal, there being no red shells far to the north of it; now you may begin to haul off, tho' you might continue safely until 7 fathoms, or in the day time even in three; that being the depth half a mile without the outer breaker: in hauling off, you will find from 10 to 12 fathoms sand & shells, and if you come near to the north east end of the shoal, you will often find live cockles, on a black sandy bottom, in 10 or 11 fathoms; your offing will be made good on a course S E b E from the first cast on red shells, till you judge yourself about 9 leagues off, then if you chuse to make *Indian River* in lat. 27° : 15 run south. But if you are bound directly to the reef, steer S 1/4 E, which will bring you in with *Hobe* inlet, where the soundings are become quite narrow. In running south from the outermost part of the shore, observe when you begin to get very coarse black sand: and black shells, for soon after you will find your bottom changed to white sand, and your water to deepen; when you find 16 fathoms on white sand, you begin to run clear of it, and in 20 or 21 fathoms, you are quite clear, then in continuing the same course your water again shoals, and in a run of about three leagues you once more find red shells, mix'd with black, and red sand, continuing a great way; you still shoal your water, and when you come in about 11 fathoms, you will have shells alone, without sand:—all this while you see no land, it making a bay to the south of the cape, till you come into about 9 fathoms, when you can just see the *Tor-*

tolas, some hills so called, in lat. 27° : 35 N. If you want to call at *Indian River* inlet, it is now time to haul in a little for the land; from the *Tortolas* southward you will again find sand, sometimes without shells; from the 9 fathoms above mentioned, it is 8 leagues to the river's mouth.

Here observe, that between this inlet, and the Cape *Canaveral*, the coast is flat, and not to be trusted so much as the shore north of the cape. Likewise observe that the red shells extend no further south than this inlet, and but a little way north of the cape shoal; whereby you have an infallible mark in coming well in upon soundings, either from the southward or northward, whereby to know that you are in with, or near to the cape shoal, if even you should chance to have been without any observation for several days.

If you intend to anchor off the river's mouth, take care to run no nearer in than 4 ½ to 5 fathoms, and chuse a spot of shelly bottom, as being safer for your cable than the sand is here.

In going from the mouth of *Indian River* to the southward, the coast is pretty bold, and your course is S S E ½ E towards *Hobe*: at near 7 leagues from the inlet we meet with some high rocks on the edge of the beach, which are an excellent mark whereby to know this place; and 4 ½ leagues further is *Hobe* or the Southern entrance of St. *Lucia* sound, this coast is so particularly described that further directions to the *Fowey Rocks* are almost unnecessary. I will however make a few remarks upon the soundings, &c. When you begin to see the above rocks of St. Lucia, you will also perceive a change in the colour of your water, the soundings are here no more then about 5 or 6 leagues broad, and I have been at an anchor in 12 or 13 fathoms water, 3 leagues from the shore, when upon trying the current with our log it was found to set to the NE at the rate of 3 ½ miles per hour, and our water was almost blue in that depth, therefore it is best to keep the shore on board, at least exceed not 2 leagues from it, your sounding will be various on sand, green owze, and shells, as marked in the map; avoid the little reef of *Grooper Hill* a little north of the *spring in the rock*, which spring is in lat. 26° 46. N. now it is necessary to keep close along the shore, within the distance of from half a mile to a mile, and you will find deep water close in; the bottom in many places coral and gravel, sometimes rocks, and notwithstanding the foulness of the ground it is yet highly necessary to come to an anchor if it should fall calm, or else you will loose as much ground in three hours calm as may cost you three days to recover.

In lat. 26° : 5 is a reef of rocks near the shore, having from 12 to 20 feet of water on it, this reef you must give a birth, and at its south end you will find the mouth of *Rio Del Medio* nearly in lat. 26, from the north end runs off a bank of soundings 5 leagues broad, reaching to the *Fowey Rocks;* this reef and bank I have before observed is in my opinion the beginning of the *Martyr reef;* the soundings on this bank vary from 2 to 25 fathoms, according to your distance from the shore, and the bottom is chiefly sand; about 2 ½ leagues S of *Rio Del Medio* is a pretty broad ledge of firm rocks joining the beach under water, which continues to *Boca Ratones,* and is not to be trusted; abreast of this ledge you will meet with many spots of coral, spunges, rocks, and grass, which through the clearness of the water look frightful to the stranger, but have no where less than 3 fathoms water on them; here fish is in great plenty.

When you are got south of *Boca Rattones* the bottom is again clear sand, and now observe to give *Fools Cape* and *Bear Cut* a good birth, on account of some flat bars and heads, when you are abreast *Bear Cut,* keep *Key Biscay* nearest on board—take care to avoid the flat off of it, which runs about 1 ½ mile off; your course is S or S b E till you are abreast of the Soldiers, & as far as them you will at least have 15 foot on a clear white sandy bottom; *Key Biscay* affords venison, and some times bear, likewise racoons and doves. It is mentioned at the beginning of this tract.

DIRECTIONS FOR THE NAVIGATION WITHIN THE FLORIDA REEF.

The next keys to *La Parida y Su Figuelo,* or, as we stile them, the *Soldiers Keys,* are 7 rocks called *Mascaras;*† the meaning of which word is, that they are but just above water. There are some mangrove and blackwood bushes on them. Their trenching is nearly to S S W. Next to these is a small island, on which are two small hills, whence the Spaniards have called it *Las Tetas,* or the *Paps;* but the Providence people have stiled it *Saunders's Key,* and the inlet to the south of it *Saunders's Cut,* where a small vessel drawing about 4

---

†This word has been in all the English maps ill copied, and transformed into *Mucares,* a sound without meaning.

foot, may enter into the wide sound between these keys and the watering places on the main.

Near 5 miles S S W of this is an inlet called Black Caesars Creek, made by the south end of Ledbury* Key; this will likewise admit small craft into the above sound.

Next is a small key which does not exceed two miles in length, at its south end is an inlet known by the name of Angel Fish Creek; to the south of this are two very small keys, and the next is an island about 5 miles in length, to which no particular name has yet been given. At the south end of which is again a very small shallow inlet, and from this the course and distance to Sound Point is directly south 2 leagues. Sound Point is the eastermost part of the land called Key Largo, though in reality it is no island, but a peninsula joined by a narrow isthmus to the main, as the maps accurately shew. This place was till 1769, always taken for an island, and together with all the keys from *Las Tetas,* and the two next to the S W of this peninsula, been known by the Spaniards under the appellations of *Cayo De doze Leggas* and *Cayo Largo,* which last name we have adopted, and by it, it is known to every one.

The general rule to sail within the reef from the *Soldiers Key* to the southward is, to have a careful man at masthead to look out, he will see all the heads and other shoals a good way off in a clear day, at least a mile; thus making the eye your pilot, come no nearer to the *Soldiers Key* than 12 foot, nor no farther E from them than 18 foot. About E S E a mile outside of *Saunders's Cut* lies a small round bank with only 9 foot on it, from this place to *Black Caesar's Creek,* the sunken heads are very frequent, and the bar of said Creek reaches a great way out. A little north of this inlet the snow *Ledbury* was burnt, after a vast deal of trouble, expence and pains taken to get her off. Right a breast of this spot, and north of the bar, is very fine anchorage in 12 foot close to the back of the reef, which makes the inlet aforesaid. On the point of this reef lies the remains of the ship *Hubbard,* cast away in 1772.

From *Saunders's Cut* to *Sound Point* there is only 11 foot of water to be depended on if you keep in that part of the channel which is clearest of rocks: you may carry more by going out further towards the reef: but the care necessary to be taken for the avoiding of the

---

*Ledbury Key, so called from the snow *Ledbury,* commanded by Capt. *John Lorain,* cast away here Anno 1760.

heads, is inconceivable. The bottom to *Black Caesar's Creek* is sand, but from thence to the S W ward is gradually changes into a kind of soft marl of the consistence of dough. When you are well clear of *Angel Fish Creek* to the south, the same rule of keeping within 18 and 12 feet depth for the channel is to be observed; but after all a careful inspection of my charts, together with a comparison of them with the course of the land you sail by, and especially a good look out, will constitute you a better pilot than any directions that can be given for this navigation.

The latitude of *Sound Point* is 25° : 11' ascertained by several good observations. It is needless to mention that the long disputed situation of *Cape Florida,* is obviously fixed in this point; any mariner that sails along this coast, will by the smallest attention to the changes of his course, find that it is the only spot that forms a true promontory, from the *Spring in the Rock* in lat. 26° 46, down to here, in the manner as my charts shew.

It is on this promontory that almost every vessel that is cast away meets her fate, I mean on the large reef that lies before it; the south point of which is dry, and forms a deep channel to go in towards two little keys called *Rodriguez* and *Tabona;* and its north point extends as far north as Angel Fish Creek. The people who watch those misfortunes to make a benefit by them, know this so well that during the summer months, the season for the return of ships from *Jamaica,* they station themselves at anchor a little south of the point from whence they can with certainty wait for the sight of any ship that is unhappy enough to be drove on shore on this reef.

Since I have mentioned these people, I cannot forbear taking notice of the abuse generally thrown upon them very undeservedly. I know of more instances than one, where they have been ill treated for their services. What the behaviour of the wreckers is in the windward passage, I cannot determine; but I appeal to every candid man, who has been so unfortunate as to have occasion for the help of the turtlers and wood-cutters, who frequent this coast from *Providence,* whether they have not always yielded their assistance with greater expedition and regularity, and with more disinterestedness than could be expected? And as for the idle tale which we are told of their making false lights on the shore, I can, from many years experience, assure it to be an untruth. Those fires are occasioned by the hunters and timber-cutters, who burn the woods to clear them of under-wood, and to procure fresh pasture for the deer. Lightning also often sets fire to trees; and I have frequently, in very dry seasons, seen spontaneous fires arise in marshy places. But after all what business has a mariner (who knows the course he must steer) to follow any light out of that course? And I would just hint to every one who passes along this coast, that on seeing a light to the westward, it behoves him to look out for breakers if he stands in for that quarter.

This promontory is a *peninsula*, though it has till lately past for an island. Anno 1769, I with great labour, fatigue, the inconvenience of musketoes, and a total want of fresh water for 4 or 5 days, explored its inside waters. I was stimulated to this enterprise by the reports of the *Providence* and *Spanish* fishermen, who told me unanimously that they had often essayed to enter at *Angel Fish Creek*, and come out at *Boca Herrera*, the creek opposite *Key Tabona*, or the contrary, but always in vain; nor did any of them know an instance of its being done. I then entered at *Angel Fish Creek*, and after a great deal of time spent and no passage found. I at last entered into the lake specified in the chart, by drawing the boat over dry ground for above six times her own length. Out of this lake I found my way by a very narrow passage at its south end; but as no part of *Key Largo* yields any fresh water, and after we got into the above named lake, all the ground around us was mangrove swamp, we were unable to find any; two of my people were nearly exhausted and spent by thirst, which we could not possibly alleviate until we reached the watering place at *Matacombe*. This by way of caution, if any stranger might come here. And after finding abundance of fresh water on every part of the coast, he may not venture to be out for any purpose, within this *peninsula* for above half a day, without a store of that invaluable necessary.

From *Cape Florida*, or *Sound-Point* to *Key-Rodriguez*, the course and distance is S W six miles; there is a good harbour for small craft to the N W part of the island, made by a reef running off from its N E end, and another good sheltering place to the S W of it, but neither has a greater depth than 9 feet at low water. This island is only a large thicket of mangrove, without a foot of dry soil on it, and affords only some *Aquatic* birds and their eggs.

Between *Sound Point* or *Cape Florida*, (by which last name I shall henceforth call it,) and this last island is another sheltering place, or roadstead, for small vessels, within a ridge or reef on which we generally seen some turtle crawls, but it is seldom occupied except by the timber-cutters, the *peninsula* affords in this place *Lignum Vitae, Mastick* and *Mahogany,* the two last are indeed found on every part thereof, but on none of the keys north of *Saunders-Key* or *Las Tetas* nor on none to the south of the last key north of *young-Matacombe,* all these timbers are however now nearly cut off.

This *peninsula* affords no living creature except *racoons* and insects, especially those troublesome ones, *musketoes* and *scorpions.*

From the *Cape* or *Sound Point,* the direction of the coast alters to S W; and the distance to the southern extremity of this *peninsula* is 10 or 11 miles. Besides *Key Rodriguez* we meet with *Key Tabona*, 2½ miles from Rodriguez, on a S W course; in going S W ward from the *Cape,* the same rule as above directed for keeping within 18 and without of 12 foot, must be observed. The channel here is pretty wide; but a man must be kept at mast head to discover heads, as some rocks lie in this tract; especially near *Key Tabona.* This *Key Tabona* has just such a small harbour within a reef as *Key Rodriguez.* This key has little or no high ground on it; it affords land crabs, a few doves, and other birds. One mile to the west of this key the south point of the *peninsula* makes a creek scarce a musket shot wide admitting only of boats: This creek is by the Spaniards called *Boca Herrera;* the bay within abounds with *Red Drum* and some other fish, and a great deal of green and loggerhead turtle; lobsters likewise are in vast plenty between the *cape* and this creek.

Within and between the above two keys lies a third called *Dove-Key*, it is very small, gravelly and rather high, with very few bushes on it; during rainy seasons it affords good fresh water and a few doves: we also find purslain growing on it. From *Key Tabona* due east lies the southern point of the great reef making a wide channel, this south point is dry, the channel is 5 fathoms deep, and ships in distress may find shelter under this point; it is almost impossible to point out any leading mark as the reef lies 10 or 11 miles from the land; but my chart shews it plain, and a good look out for heads is the most essential pilotage here.

From the southwest part of the *Peninsula* the course is S W distance 6 or 7 miles, past two more keys of no note, still belonging to the *Groupe* of isles called *Key Largo,* we then come to an island of some note called *Young Matacombe,* it is 4 miles long, trending S W;

it has nothing remarkable on its except a well of good fresh water on its east end, this however is known to few and consequently little frequented; off the S W end of this key lies a small drowned mangrove island, and a channel 10 foot deep runs in to the south of it and shoots up to within the great key, but there being nothing on the key to attract much notice it is seldom visited.

In coming this way from *Key Tabona* the channel is in general deeper than before, but the same rule of keeping without 12 and within 18 feet still holds good; directly abreast of *Young Matacombe*, 1½ miles E from the land are a parcel of dangerous sunken heads called the *Hen and Chickens*, which require a good look out to be kept.

Next is *Old Matacombe*, remarkable for being the most handy and best watering place on all this coast, on its east end are 5 wells in the solid rock, said to be cut by the savages, but to me they appear natural chasms, they yield excellent water in abundance, and some ponds near it likewise afford some, insomuch that in a wet season all the east end of the key is overflown, and water enough to be had for to supply the necessity of a fleet; the key is 4 miles long, and at its west end are likewise some ponds and wells, but the water is of a much inferior quality. This key was one of the last habitations of the savages of the *Coloosa* nation: about a mile to the east of the N E end of *Matacombe* lies a small bushy, gravelly key on the extremity of a reef, this key is called *Matanca*, i.e. Murder, from the catastrophe of a French crew said to have amounted to near three hundred men, who were unfortunate enough to fall into the hands of the *Coloosas*, which savages destroyed them to a man on this spot. This *Key Matanca* lies SS W 2 miles from the small man-grove key & is the leading mark for finding the watering place on *Matacombe*, for come from whatever quarter you please, the way is to run boldly up to within pistol shot of the N E part of this small key, when you will find yourself at the chops of a channel which leads up to the watering place; it admits of 10 foot and is bounded on its N side by a very shallow reef which divides this channel from the channel to the south of *Young-Matacombe*, the channel is so plain that the best direction is the eye; the tides are very rapid, consequently require to be consulted both in coming in or going out, the channel is very narrow having only just room enough for a small vessel to turn to windward.

From the south end of the great reef of *Tabona* it is no longer a continued reef, but all in divided spots as the chart shews, and the channel within them is likewise wider and deeper, but you will find less smooth water than from *Key Biscay* to *Key Tabona*.

To the west end of *Matacombe* is another channel which is the first that opens (for a vessel of 6 foot draught) into the gulph of Mexico.

To sail into this channel you must borrow pretty near to the key at its west end, where you will have 10 foot water, then steer in north towards a cluster of small mangrove keys for about a league; when you are within half a mile of the bank of said keys you alter your course to W N W and you will soon find yourself between two very shallow banks, and shoal your water to 6 foot; now you will see many points of banks which will oblige you frequently to loof and bear away between the W, W SW, & W N W, the banks are almost dry, but the channel of an even depth shows blue; so that a careful eye cannot miss it. These courses continue for 4 leagues, when you will find your channel to widen very much. Now steer N N W 3 miles and you will see *Cayo Axi*, or *Sandy Key* in the N W about 1½ leagues off; then alter your course to west, come no nearer than within a league of the key, and in five miles you will deepen your water to 9 feet, continue your course W till the key bears E N E and you have sea room into the *Gulph of Mexico*, but the soundings deepen very slowly to the westward and northward; take care never to attempt this passage without a leading wind.

N. B. *Sandy Key* is a low key covered with some bushes and prickly pears; there are rabbits on it, and always plenty of *Flamingoes, plovers*, and other excellent water fowl to be had.

N W four miles from this key is *Punto Tancha*, or *Sandy Point;* here was anciently a settlement of *Coloosa* savages. Tolerable water, and excellent venison are to be had here.

From the S W end of *Matacombe* to the west end of *Cayo Vivora*, the course is S W, and the distance 4 miles, the depth of water 17 or 18 feet on a sandy bottom, give the key a birth of at least 2 miles. At the S W end is a channel of 9 foot depth into the Gulph of Mexico, plainly marked on the map, for which a good and careful look-out is the best director.

From *Cayo Vivoras* S W b W 10 miles brings you to a remarkable

---

*Spanish for Rattle-Snake Key.

contraction of the channel between the outer reef and the keys, your depth is generally 18 foot, the bottom sandy, and a considerable broad bank runs off from the islands, which are the *Cayos Vacas,* i.e. *Cow Keys.* At this contraction of the channel your course alters to W S W, and you go through the like draught of water for 5 leagues, then the reefs are again further off, and you are come to the westermost of the *Cow Keys,* in lat. 24°: 41, where another channel of 6 or 7 foot deep goes into the *Mexican Gulph.* On these keys is tolerable water and plenty of deer.

The next islands are called the *Bahia Honda* keys, extending E and W for 5 leagues, your course is W : and your depth of water 17 or 18 feet; at the east end of the westermost key is another channel 8 feet deep, opening the gulph of *Mexico,* and leads as the chart explains; but being difficult, it is seldom attempted. Midway before these islands 2 leagues to the south is a dry sand-bar, on which, in the season, the sea-fowl lay a great quantity of eggs. Five leagues W b S from this, and directly south from new-found harbour, 3 leagues off is another, on which his majesty's ship the *Loo* was lost, from which accident it has acquired the name of *Key-Loo.* From here to **Cayo-Hueso,* vulgarly called *Key-West,* the course is W b S, and the distance about 6½ leagues. At a little more than 2 leagues you find a harbour called *Newfound-Harbour;* it is safe, but the bottom foul in many places. The depth of water in this course is from 17 to 23 feet. About half way between *Newfound-Harbour* and *Key West,* are some heads of rocks, as laid down in the chart, but they are easily discovered in fair weather. All these keys abound in venison and water; especially the *Pine Keys,* on which there is likewise some honey found.

At the west end of *Cayo Hueso,* is a passage into the *Gulph of Mexico,* this key, or rather these keys (being made up of several) extend about SW and NE for 10 miles, having a shallow bank before it. The SW end lies in lat. 24°. N long. 81°: 15 west; it has a shallow ledge close to it on the south, and the point is a low kind of savannah, on which one or two single trees appear; near which is a well of very ordinary water. The way into the channel is to give the south point a birth till you have the channel well open to the north of you, then steer in and keep the key close on board within 15 or 20 fathoms, you will find from 24 to 16 foot water, run in till you bring the NE

*Spanish for *Bent Key.*

end of a shoal on your larboard hand to bear near WNW, then double the point of the island to the westward, and you may anchor in 16 or 18 foot water, but the bottom is foul and the roadstead unsafe, especially in N and NW winds. Round the shoal to the westward is another channel rather deeper, but not so direct; and half a mile to the north of the shoal is a bar having scarce 14 foot on it; after passing which you may steer N ½ W and you will in a run of 2½ leagues clear the eastern bank, gradually deepening your water, as the map shews. The tide here runs violently. Within the anchoring place eastward of the point, in a path leading to a well or pond of excellent fresh water; round which a low kind of stone wall is placed, and the trees are marked with many names; the ground is trodden like a sheep crawl, occasioned by the deer who resort here to drink, of which a patient man may here shoot 5 or 6 in a day; they are very small, but of excellent flavor. Doves likewise abound here. This watering place is ¼ of a mile from the beech.

If you are bound to sea from this key, sail from the SW end directly for a sandy key, which you wlil see in the SE near 3 leagues off, and you will have generally about 23 or 24 foot depth, borrow close on the west end of this sand so as you might chuck a biscuit on shore, and you will find 5 fathoms, which suddenly deepens, and two miles carry you off of the bank into the gulph stream. 3 leagues west from this sandy key lives another similar to it, being the last part of the reef, the *Havana* bears from here nearly S ½ E, and is about 20 leagues off; but vessels bound there will do well in steering higher up, if possible, on account of the currents. The tide ebbs & flows here regularly 6 foot, and the time of full sea at full and change of the moon is 8 o'clock, as it is every where from Key *Vacas* to the *Dry Tortugas*, the tides setting as the darts describe. The tides from Key *Vacas* northeastward rises not quite so high, and the time of full sea is from 7 to 8 o'clock, being later as you come westward; the darts shew the setting of the tides: to the northward of *Key Biscay* the current on soundings is much governed by the wind; but when the wind has little influence the ebb sets north, and the floor southward. A due attention to this will much shorten a passage over soundings to the reef.

Next are the *Cayos Mulas*, to the westward of *Cayo Hueso*, trending east and west for 6 leagues; at whose west end we find the principal passage out of the Gulph of *Florida* into the Gulph of *Mexico*, by the Spaniards called *Boca Grande,* which I have never passed, but such as it is marked in the map, I have taken it from a Spanish manuscript, which bore more marks of genuine accuracy than most of the paltry draughts used by that nation. It was communicated to me by a gentleman whose name was D^n *Manuel Hidalgo,* a very experienced commander, who assured me that the Spanish galleons from *La Vera Cruz* in order to avoid the east winds keep far enough north to make the *Bay of Tampe,* from which they shape their course so as here to disembogue out of the Gulph of *Mexico* into that of *Florida,* passing a little east of the Key *Marques.* I have seen two of these ships on the coast of *West-Florida,* not far south of *Cape Blas;* which circumstances and that of three galleons being cast away near the bay of St. *Bernard,* seem to confirm that gentleman's information. The sounding on the outside upon the bank are taken by myself.

The bank of *Marques* extends east and west for 6 leagues, then N W ward for near 6 more, having no reef before it.

From here to the *Dry Tortugas* the course is W½S, and the distance 44 miles, to reckon from two miles south from the middle of *Key Marques;* water of various depths: the southern part of the rocks lie in lat. 24. 24. but a round bank with about 6½ fathom on it lies before it in lat. 24. 18°: the longitude is 82°: 18 west from London. In August of the year 1766, I was becalmed near this place, and drifted for 3 days, not having had an observation for 5 days before on account of fogs. I was very uneasy especially on the appearance of a dove on the evening of the last day, this kept us all night (more than common) watchful; and at the break of day next morning we sounded in 30 fathom. We kept the lead going, and shoaled our water gradually to 20 fathoms, when it was full day, but still foggy, on which I gave orders if the water shoaled to 12 fathom to come to an anchor (our drift was then with the current, which I afterwards found to be the tide of flood setting to the northward on the bank) and I left the deck; but had scarce sat down in the cabin when the vessel which drew between 8 and 9 foot water struck, and as it were dragged over a coral bank, I started up and hove the lead overboard, when I found 18 foot water on sand, and saw a dry bank or key close too to westward, the current still running with violence. We came immediately to anchor, and threw the boat out, in order to examine the place where we were, we found ourselves surrounded by three very small low sand keys (full of prickly pears) on the west, and a reef every where else. About 9 o'clock the tide began to ebb, setting violently over the reef to S E of us; at 10 the weather became clear, and we saw bushy islands to the N & N N W about 4½ leagues off. I had a good observation at noon, and found the lat 24°. 25. The two following days we were employed in looking for a passage out, through which on the morning of the third day we warp'd out to the east. It was on the day following the full moon in August when we struck, we observed the tide to rise and fall full 6 foot, and the place where we struck at first to have between 6 and 7 feet of water on it, when the tide was out, it being 9 o'clock when it began to ebb; I thence fixed the full sea at about 8 o'clock on the full or change or a S E b E ½ moon. All these observations I have since been confirmed in by a more frequent experience. This adventure was my first knowledge of the *Martyrs*, and was the raising of the first idea in me of publishing this work. The calms continuing I tided it up within the reef as far as *Matacombe*, where we watered and proceeded out into the stream; I became very sensible of what a great benefit the knowledge of this reef is to all people, who by means of calms, or otherwise have a long passage from *Cape Antonio* up; especially on account of refreshing. I was then bound from the *Isle of Pines* to *Georgia*, with mahogany on board.

The soundings, as I have marked them, were obtained at several times passing near and around this shoal, they are remarkable, and for them you may rely on the draught. I observed the variation of the compass to be 5° : 47′ east.

On the N W corner of the bank is a harbour where I have made the observations that are laid down in the draught, at three or four different times; but as it is only planned by the eye, never having had an opportunity to make a circumstantial survey I shall forbear saying any more about it than that the northermost part of the bank lies by good and repeated observations taken by myself in lat. 24° 48. N.

Before I proceed into the *Gulph of Mexico* I shall say something more concerning the reef, and give my advice to such as are bound through the gulph of Florida.

Many people talk about going southward thro' the *Gulph of Florida* by keeping upon soundings outside the reef, but it is a navigation so dangerous, that I would wish every body (except in cases of necessity) to avoid it. I will however, transcribe the remarks of

Robert Bishop on this navigation, and add a few observations of my own on them.

"'When we were sailed out of *Port Royal Harbour*, we kept upon soundings, till we came as far to the southward as *Tybee*, and then we stood to the E and afterwards to the S till we were got into the lat. of 26 d N and then run down in that parallel, and made the S end of *Abaco*, or *The hole in the wall*, and N W by W dist. 5 or 6 miles from it, we anchored in white water, of about 7 fathom, off a point where they key falls in, & there we got fish in plenty, for which purpose we anchored, as also to delay time for 4 or 5 hours. At 2 P. M. we weighed, and stood S W for the *Berry Islands*, which we made and came upon soundings at 8 ditto. The course is S W distance 8 leagues; and from those islands to *Providence*, the course is S S E distance 12 leagues.

"When we got soundings, we kept our lead going, & we laid by it all night, keeping in, or out, as we deepened or shallowed our water; but our course was between the W and W by N and our dist. 15 leagues.

"The breadth of soundings (at least from the *Berry* to *Isaac Rock*) is from 3 to 5 miles from the banks, being broadest at the rock. Upon the bank's edge are two small rocks, between the *Berry Island* and *Isaac Rock*, which terminates the N W corner of the bank.

"At a cables length off the W side of *Isaac Rock*, you have no soundings: So that as you come round it, you immediately get into ocean water, and consequently into the current of the *Gulph*. But if it should happen that the wind be at S S E and you cannot lie S or S by E so as to take the current under your lee, and keep the bank on board, your best way will be to anchor, or keep in upon soundings to the northward of *Isaac Rock*, till the wind comes favourable.

"By this will appear the necessity of getting to *Isaac Rock* by morning; for then you get round and keep in on the edge of the sounding, by which you will plainly discern the bank, the water being clear, and the bank white, with two small rocks between *Isaac Rock* and the island of *Bimina;* the distance between being 4 or 5 leagues almost N and S.

"At *Bimina* there is a harbour of 9 feet water, and anchorage in the opening, with a well of water on the E point. The harbour lies in lat. 25 d. 30 m. N.

"From the island of *Bimina* S. dist. 8 or 9 leagues, is *Cat Key Harbour*, or the beginning of the *Rocqueses*. From *Bimina* the sounding is narrow, and, consequently the current strong. From hence over to the *Florida* shore, the breadth is no more than 15 or 16 leagues. From *Cat Key* abreast of the *Rocqueses*, the sounding is pretty broad, with good anchorage, and less current, as you come to the S E and S E b S.

"From the second or *Cat Key*, the course S E by S dist. 12 or 13 leagues, is *Orange Key*, having good anchorage to the S W of it 5 or 6 miles, in 20 fathoms water. When we get upon this flat, we think we have secured our passage through the *Gulph*, this way; for then you may make sail either in the morning, or at midnight, steering S W dist. 10 11 leagues, and so you will fall in with *Key Sel Bank*, which for ten leagues on the north side stretches E and W and consequently the current sets stronger as you come to the westward. When you come over, there is good soundings all along by it, and you may discern by the bank how far you are to the eastward of the *Double-Headed Shot:* For as my draught shews, the number of rocks on the bank, there is anchorage by spots all the way in; but the soundings are narrow at the *Double-headed Shot*, the middle of which lies in lat. 23 w 57 m N and S S E is *Key Sel*, where there is water, distance 4 leagues. Here the *Spaniards* make salt.

"From the *Double-headed Shot* to the edge of the soundings a little to the W of *Cape Florida*, is N W by N 15 leagues.

"From the *Double-headed Shot* to the *Metances*, the course is S W by W dist. 23 or 24 leagues, and from the *Metances* to the *Havanna* W much the same distance; off of which we cruised 5 or 6 weeks, and almost every day gave chace. The first that gave us a jaunt for the *Gulph*, was a ship from *Jamaica* bound for *South Carolina;* We followed her in a dismal dark night, and at one or two in the morning had like to have run her down. There was a constant order on board, to heave the lead every half hour, the whole cruize; and when we brought the ship to, we had 70 fathom water. As our pilots were no way concerned, we lay there all night, and fell off to 100 fathom in the morning; at day-light we could but just see the bushes of the key off the cape from the poop, and it falling calm, we drove off soundings, and so got into the current, which carried us as far as 25 d 30 m N when the easterly wind sprung up, and we went into soundings and out of the current; so hoisted our boats out, and made them keep by the outside of the reef. Whilst the ship went upon the edge of soundings, we did not only know the breadth of sounding by the distance the boats were from the ship, but they besides supplyed the ship's company with fish in great plenty, and so for 23 leagues, 15 of which to the N E of *Cape Florida,* and 8 to the W S W till at last we arrived at a sandy key, where the *Looe* was cast away. At that key is the going into the inside of the reef, or the channel, through which all the *Spanish* vessels pass, that go from the *Havanna* to *St. Augustine.* In this channel there is 4 and 5 fathom water, but in some particular place it is shallower. The channel continues as far as lat. 26 d N or as far as the reef runs. Off *Cape Florida,* is the broadest sounding, where we met with 100 fathom, with the bushes just in sight from the poop of a 40 gun ship, and distant from the reef 6 or 7 miles. But in lat. 25 d 30 m are other rocks, where the *Fowey* was lost: At this place there is sounding two hawsers length off the reef in 30 fathom; but one mile further it is not so broad; so that I judge the *Looe Key* to be the beginning of the sounding off *Cape Florida,* and the *Fowey Rocks* the ending.

"In consequence of the foregoing observations, when we chaced we had no difficulty in going back. If we chanced to chace but as far as 25 d 25 m N we went immediately in upon sounding, and up to *Looe Key.* Then we could be off the *Havanna* the next day in our station; so that now we had not near so much trouble as we had at our first coming on our station; for then between *Isaacs Rock* and the two *Keys* of the *Rocqueses,* we were always afraid of driving quite through the *Gulph;* but on this side, if we could but keep in upon sounding, we were safe.

"Indeed most men who have been through the *Gulph,* when they come so far to the northward as 24 d 30 m N must be sensible that they meet with a strong current, with the rippleing and boiling of the water; which current sets between the N E and E N E and is occasioned by the edge of sounding that comes from *Looe Key.*

"The Florida shore does not go north as has been formerly imagined, till you arrive in the lat of 25 d 40 m N I could hartily wish that the *Old Streights of Bahama* were thoroughly rummaged; for by what I have seen and heard, it is not so bad a navagation as many now think it to be. I believe it to be an easy way to go to the *Missisippi,* by crossing over to the *Florida* shore, and so round the *Tortudas Bank.*"

These remarks of that gentleman are very judicious, and indeed most of his performance tho' strangely unconnected is generally pretty just, but his charts bad. I will however (after asking his pardon) endeavour to rectify a few of his annotations, which seem rather crude or too hastily penned down.—

The S W course which he mentions from the *Hole in the Rock,* will carry you to the *Berry Islands,* as he says, but so low down as below *Little Harbour* which is at least 6 leagues too far to the S E. I would advise running W b S or at most W S W half W 13 or 14 leagues, which will bring you to *Stirrup's Key* or *Money Key* being the N W part of the *Berry Islands.*

To the westward of *Stirrup's Key* is a tongue of ocean water shooting into the bank, across which your course is W half N or W b N for 10 leagues scarce, to the little *Isaacs Rocks;* which rocks I suppose he means when he mentions two little rocks between *Berry Island* and *Isaacs Rocks,* from these to great *Isaacs Rocks,* the course is about W and the distance between 9 and 10 leagues.

What he says of great *Isaacs Rocks* and the getting round them is very just; but the distance from great *Isaacs Rocks* to *Bemini* is near 7 leagues. Mr. Bishop says but 4 or 5.

The harbour of Bemini is well described. From *Bemini* to *Beaks Key* is S half E 7 1-2 leagues. *Beaks Key* is the southermost of the *Cat Keys* and affords some shelter.

A little less than a league from *Beaks Key* to begin the *Riding Rocks,* being rocks bare of bushes, and looking like wrecks, the Spaniards call them *Los Membros.*

S W b S about 7 miles from the southermost *Riding Rock* is a shoal on which a very rich Spanish galloon struck in 1765, her bottom beat over, and pieces of her were found every where as far as *Money Key* to the eastward; the loss of this vessel was a *profitable ill luck* to the people of *Providence,* the bottom lies now about E or E b S 7 or 8 miles from the shoal in 17 feet water, and is yet supposed to contain some treasure.

S S E from the shoal 3½ leagues is *Orange Key* & the *Rocquesses,* here the anchorage &c. is as Mr. *Bishop* describes them, and his directions to *Deadmans Keys,* or the west end of the *Double-headed Shot,* are to be depended upon. His relation of going southward upon soundings near the reef is very just, and the method of keeping a boat on the reef a very proper one; but this is too dangerous a navigation to be attempted by most people, yet if a gale should come on from the eastward, I do not apprehend the risk to be so great as might be at the first thought imagined, nor to be so much dreaded as by a person coming from the southward and bound through the Gulph, who must needs be more uncertain of his ships place, than him that is bound southward on soundings, for the nearness of the very rapid current of the gulph, will always enable the latter to take the proper precautions for clawing off with certainty of success.

The distance from the *Fowey* rocks to *Cape Florida* is 15 leagues, as Mr. *Bishop* observes, but he makes an egregious mistake in calling it only 8 from the *Cape* to *Key Loo;* for upon supposition that the south end of the reef before the *Matacombe* islands, be the true cape, (which indeed it is of the reef, though not of the land) the distance from there to *Key Loo* is above 20 leagues on a W S W course; and from the south part of the true *Cape Reef* to the south end of *Matacombe Reef* is more than 4 leagues more; I therefore guess him to have been misinformed with respect to the local situation of the loss of the *Loo,* and that he mistook the small dry sand-bar upon a reef before *Key Vacas* for *Key Loo;* this sand-bar being the first on the reef, and lies 11 leagues to windward of *Key Loo.*

There are five of these dry sandy islands of which *Key Loo* is the middlemost.

There are many inlets into the reef, but to attempt describing them each apart, would be to throw more obscurity on the matter, and to open a way to more danger, which has already been too much done by the absurd author of that paltry pamphlet, called the *Atlantic Pilot*. I shall content myself by telling my reader, that the necessary precaution of keeping a boat on the reef, will always point out to him these cuts in such a manner that he may safely enter any one of them in moderate weather, if want of water, contrary wind, or any thing else make it necessary for him to take shelter under the reef. Two of these entrances however require a little more to be said of them, one is at the south end of the *Cape Reef* in lat. 25° 2' directly east of *Key Tabona**, the other at the south end of *Matacombe* reef; E S E from the north east end of Matacombe in lat. 24° 52': The first has got a dry knowl of rocks above water on the S E point of the reef, directly on the edge of the channel, whereby it is easily known;

*Tabona is Spanish for a whang or horsefly.

the other has no such visible marks, but the eye will guide you for both, especially the northern one, where the land may also help a little, as the two small mangrove keys *Tabona* and *Rodriguez*, lie a good way from the land, nearer to the reef, and consequently shew themselves plainer in the west; for the rest all the land appears so much alike, that it requires years of experience to learn to know it. The soundings in &c. are as my chart directs, and I need not to any person (who knows that in a gale by reason of a reverting current, anchoring is full as safe under a reef as under land) to enlarge much about the utility of the knowledge of the channels, much less to a man who is either in want of water, or who upon falling in with these shoals and thinks himself in danger, has manly courage enough not to suffer his fright to overcome his reason: And whoever happens to be overtaken by the vehemence of gales, when too near in upon these reefs, may be happy in knowing that there are such entrances into safety.

Mr. *Bishop* mentions the depth within the reef at 4 or 5 fathom, which is more than double the quantity found. This is, I suppose, an error of the press.

N. B. Wherever I have occasion to mention soundings, they are meant at *spring-tide, low-water;* and all the soundings are laid down on my maps in feet, to avoid confusion, occasioned by marking feet in one place, and fathoms in another. The Roman capital figures denote the time of full sea at the respective places (where they are marked) on the full or change of the moon.

The quantity of fish and turtle to be caught here is really amazing, which, joined to the many watering places, with the plenty of venison, and bear-meat, make this coast a valuable *rendezvous* for cruizers in time of war: spars may be had here at all times, either out of the pine woods, back of the keys, or among the drift on the beeches, which is no small inhancement of the value of the coast to such vessels; because they are not seldom in the want of them; at least they are more liable to such losses than merchantmen are.

The fish caught here are in such variety, that a bare catalogue of them would take up pages, These most commonly caught are such as seamen know by the following names, viz. *King-fish, barracoota, tarpom, bonita, cavallos, amber-fish, pampus, silver-fish, jew-fish, rock-fish, groopers, porgys, margate-fish, French margate-fish, hog-fish, angel-fish, yellowtails, red, grey* and *black snappers, dog snap-*

pers, mutton-fish, grunts, murenas or *muray, mullets, sprats, mangrove snappers, parrot-fish, red* and *black drum, bone-fish, stingrays, sharks, lobsters,* and an immense variety of others, all excellent in their kinds, and what renders this plenty of more worth is, that we may with safety eat of all fish caught on the *Florida* shore, unless it should be of hog-fish taken on the very outer reef; for I have heard of an instance of one of this kind having sickened some people, but of my own experience I can say that I have always eaten that delicate fish with safety, and even the *amber-fish,* and *yellow billed sprats.* The worst, or most violently deleterious fish in places where fish are poisonous, is here always eaten with safety: On the contrary, on the *Bahama Banks* it is requisite to be cautious what fish we eat before trying, which is most conveniently done by cutting the heart out of the fish as soon as caught, and to bite in it, when if the fish be bad, it will leave a very nauseating, bitter, astringent taste on the tongue; but if good, no such taste will be perceived. The method of boiling silver with the fish is not so certainly to be depended upon.

I judge now that I have said enough about this mazy navigation; and will therefore proceed to give some

DIRECTIONS FOR COMING ROUND CAPE ST. ANTONIO, THROUGH THE GULPH OF FLORIDA

It is sufficiently known that from the south point *Negril* in *Jamaica* to the *Grand Cayman* Island, the course is W N W somewhat westerly, and the distance 53 leagues, that off the S W point of the island a ledge of rocks extends almost a league into the sea, and from said island to Cape St. *Anthony* the course is N W b W 87 leagues, but it is adviseable to take rather a little more westing for fear of falling in with the *island of Pines* or the *Jardins,* which reach eastward and southward, or more properly on a course E by S half S from said *island of Pines* for 25 leagues, and are dangerous; should you fall in with this *island,* its shore to the southward may be made pretty free with, the course and distance from the *Grand Cayman* to this is N W somewhat westerly, 50 leagues, and from the S W end of the *Isle of Pines* to *Cape Corientas* is W half N 24 or 25 leagues; at the S W point of the *Isle of Pines* I once put in for wood and water, and found mahogany growing so handy that I took in about 4000 feet of it in a very few days; but the water was difficult to get, though good; we got it in wells in a savannah at some fishermen's huts, near the very S W pitch of the point; this place is however only fit for small vessels to call at: there is a Bay to the northward of where we lay, in which I was told that a channel of 18 or 20 feet depth might be depended on; we got plenty of fish and turtle while we stayed here: off the S W end of the island are two very small keys, which are very shoal all round.

At *Cape Corientas* is anchorage in 5 or 6 fathom about 2 miles to the N W by N or NNW of the cape, but you must be careful in going into this bay to give the cape a birth of a league at least, on account of a shoal that runs to the S W off from the pitch of the cape; the water in the wells is indifferent, but easy enough to come at.

From *Cape Corientas* to *Cape St. Anthony* the course is W a little northerly distant 12 or 13 leagues; round the cape under the W end of *Cuba* is anchorage in 5 fathom almost every where; to the north of the cape between it, and a mangrove point about 2½ leagues from it: and about half way between them are some wells of indifferent water.

If you are bound through the *Gulph* of *Florida* steer from Cape St. Anthonio N b W half W or NNW and after a run of about 5 leagues you will find yourself in 15 or 16 fathom water, the south point of the Colorados then bears E b N 5 leagues off; this point of the reef I observed in lat 22° 56 and the Cape St. *Anthonio* lies in lat. 22° 36 long. 83 52.

I could say more on this subject, but a small matter of inspection into my maps will shew that there is a vortex of current near and within these shoals: I therefore judge it recommendable to any person bound round this cape, to keep a N N W course for about 8 or 10 leagues to avoid them, and when you come as far to the northward as the lat. 24°: 00 N, by all means keep a good lookout, for in this passage you will nine times in ten meet with a strong current out of the *Mexican Gulph* setting eastward, sometimes at the rate of 2½ knots; this current will set you on soundings on the *Tortuga* bank before you are aware thereof. In foggy weather these soundings may be a guide, my charts mark them properly. You do not change the colour of your water till you get well in with the shoal, but there is generally an eddy current as soon as you are on soundings. The south part of the shoal whereon I struck, as related page 256 lies in lat. 24°: 23 N, and the southernmost dry island in lat. 24° 25. Therefore if you stand over to the *Florida* shore, as soon as you are up to the lat. 23 : 25, keep as much to the eastward as N N E or N E b N till you get soundings. And whatever terrible idea people have of that shore, if the wind will allow you, keep it on board, especially in the autumn and winter season when the N and N W winds are frequent, and the current often runs to leeward. In that season you may take an advantage of the tides on soundings, by carefully observing their times, & this conduct will tend to shorten your passage: When, however, you are got as far windward as the south end of the *Matecombe* reef, in lat. 24° : 52, long. 79° : 50 W from *London*, endeavour to get all the easting you can possibly acquire, in order to get the *Bahama* shore on board; which I shall hereafter, by pointing out the soundings, shew to be by much the most eligible for safety in going northward.

If you intend to beat up on the *Cuba* side, which is the safest on account of its bold shore, and perhaps the most expeditious in case of a weather current, (which runs here with the most amazing velocity) you had best not go further north than the lat. 24° : 00, and on your first tack you will know whether the current is favorable or not, for if you make not at least the *Bay* of *Honda* you may be certain there is a lee stream, & in that case an attempt to beat up under *Cuba* will be nothing better than kicking against the pricks, and the *Florida* shore ought to be attempted. Those unlucky persons who tell the dreadful stories of being 5 or 6 weeks, and even more, in getting from *Cape Antinio* through the Gulph of *Florida*, which is but too frequently true, have met with such a current, and through fear or ignorance of the north shore, have lain that long spell wearing and tearing vessels and rigging, expending their provisions and water, and fatiguing their men to no purpose.

When you are up to the *Coloradas*, the high land of *Punta Abatas*, or *Cape Bonavista*, shews itself remarkable, as drawn in the map : It is seen to the E 23 or 24 leagues off, which may be of use to know when you are near that dangerous reef.

The *Coloradas* are the west end of the reef of St. *Ysabella*, a steep dangerous shoal extending a length of 27 leagues from *Porto Puercos*, or *Hog-Harbour*, to the aforesaid *Coloradas*, on a course chiefly W b S, and on the edge of the reef nearly half way between these two extremes, lies an island called *Key Lavasa*, from whence you see the notch of the *Cox Combs*, nearly ESE ward 6 or 7 leagues off. At this key is anchorage. The entrance of *Porto Puercos* is readily known by two bluff islands at its mouth, and the notch of the *Saddle-Hill,* which in the charts is marked * bearing south over it. The *Table Land* and the *Saddle Hill* over it, shew as I have delineated them. When you bring it to bear about E, and the western part of *Punta Abates* about S E, you will then be within less than a league off from the reef, and in standing northward again, you will soon open the *Saddle Hill* from the *Table Land,* which is a certain mark for keeping clear of the reef of St. *Ysabella,* and may be made use of in beating up under the reef, which method is however not to be advised.

N. B. This *Table Land* is called *Mesa Maria,* but must not be confounded with another of the same name 8 leagues to the leeward of the *Havana.*

Mr. *Bishop* remarks, that keeping the Saddle Hill open with the notch of the Cox-Combs, will likewise serve as a mark for working to windward under said reef, but I never saw his performance till after I had been the last time here; thus I could not compare it on the spot, nor do I remember exactly to have taken notice of this.

He also says, that if you bring the *Saddle-Hill* on S W b W, and keep that bearing, it will lead you into the harbour of *Bahia Honda,* where he says there is a little island in the middle of the bay with a good well on it; and S E from this island is the *Rio Honda,* a fresh river. These remarks I never had an opportunity to experience, but being of importance, I have inserted them here on his credit.

This *Bahia Honda* lies near 6 leagues east of *Porto Puercos* and the *Havana* 19 leagues to windward of *Bahia Honda.*

I have given as near as I could, a representation of the land from *Havana* to *Pan de Matanca,* the paps are very easily known, and by bringing them to bear S, or S ½ E, then running for them, it will lead you into the port of *Havana.*

This place, by my observations, lies in lat. 23°: 20, and its long. I judge to be nearest 81°: west from *London.*

From the *Havana* to *Punto Ycaco** is 23 leagues, the course is E ½ S nearly. From this point we see the *Pan de Matanca* to the W S W of us.

I have always observed that a lee current does not extend eastward of *Bahia Honda,* at least I have found it so during 5 or 6 times

---

*Mr. *Bishop* calls this point *Jacko. Ycaco* is *Spanish* for *Coco Plumb,* of which fruit there is abundance on the point.

that I met the current setting westward; but at those times and in 8 or 10 other passages along this coast have always found a strong windward current east of that place, so that if I once weathered it, I was seldom more than 12 hours in getting up to *Punta Guana*, being the west point of the Bay of *Matanca*.

From this point most people chuse to stretch over, and Mr. *Bishop* advises the bringing of the *Pan de Matanca* S, or S b E, before one puts over for the gulph, and then to steer N N E; both are right enough; but I shall endeavour to point out a safer way.

From *Punta Guanos* to *Key Sal*, the course is E N E somewhat easterly 23 leagues.

From the small key of *Punta Ycaco* to *Key Sal*, is N E b E near 13 leagues.

From *Key Sal* to *Deadmans Keys*, on the *Doubleheaded-shot* bank, is N b W 5 leagues. And,

From *Deadmans Keys* to the south point of the *Cape Reef*, is N W b N 19 or 20 leagues.

From the same *Deadmans Keys* to the *Riding Rocks*, is N E b N 23 leagues.

And the *Riding Rocks* lie E b N 19 leagues from the south point of the *Cape Reef*.

Hence it follows, that from *Punta Ycaco* to the south point of the *Cape Reef*, the course is nearly N, and the distance 29 leagues.

It is hardly possible that a vessel should fall so far leeward as the sand bar of *Key Vacos*, or *Key Loo*, unless that supine carelessness and drowsy humor too common among the greatest part of the English sailors, might cause the crew to let her run 4 or 5 points leeward of her course.

For those, however, that may meet with such an accident, I will point out the situation of those keys.

Mr. *Bishop* has evidently mistaken the sand bar of *Key Vacas* for *Key Loo*, which may easily be the case without my impeaching his judgment, for they look exactly alike: I am led to think so when he says, "from the *Havana* to—*Looe Key*, is N E b N, distant 32 leagues." And again, "from the *Metances* to *Looe Key* is N westerly, distance 25 leagues." Whereas the true bearing is from the *Havana* to *Key Loo* N b E 21 leagues, and from *Punta Guanos*, or the west point of *Matanca* Bay, to the same key, the course is NNW 24 leagues. But from the *Havana* to the sand bar of *Key Vacas* is N E northerly 32 leagues. And from *Punta Guanos* to the said sand bar, it is North about 24 leagues.

From the foregoing Courses and distances the Intelligent mariner will easily see, that *Punta Ycaco* is the most eligible place from whence to take his departure when bound through the *Gulph* of *Florida*. My advice therefore is, that if he makes the keys of *Punta Ycaco* early in the day, to delay time till 3 o'clock P. M. at least, or rather 4, his time may be usefully employed in fishing on the reef north of the point; for his labour and delay will be amply compensated by the excellent refreshment and store of fine fish it will afford him. The depth of water is from 4 to 14 fathom; but venture no nearer than 4 fathom, and beware of a sunken rock lying about a mile westward from the westernmost key.

By doing this, and towards evening taking your departure from here, & steering NbE, or rather NNE, if you are to see land on the western shore, you will see it early in the morning, (provided you have wind enough) & thus your safety through is secured, but by all means endeavour to get as much easting as you can, (unless in winter when the winds hang to the westward, for in that case *Florida* is safest) to get next to the *Bahama Bank*, for here you find pretty broad regular soundings, and have a weather shore, whereas *Florida* has little or no soundings and is a lee shore. For nine tenths of the year, you may venture on the *Bahama Bank* always safely as far as in 6 or 7 fathom, which circumstances will acquaint you of your safety in the night, should you be more than one day getting through, and be on the weather side of the passage: Let me however caution you to get off the bank before you pass the lat. 25° : 30' to the N ward for fear of *Isaacs Rocks*.

During my several cruizes within the *Martyr* reef, I have seen a great number of vessels borrow so close on the reef, as that they appeared to be within it, and sometimes I could even see the people with help of a glass, such I suppose are well acquainted, or very bold: but be the man who does this never so experienced, he must be careful to keep a strict look-out; for my part I would not come nearer than just to raise the land, especially as tides may have their influence further off than we are aware of.*

*See the end of this appendix.

The *Florida* shore does not run north till you are past the *Fowey* rocks in lat. 25° : 35 N; to say more about it would be no better than needless repetition, as I have already so amply treated of it. But as most people, even at this day, imagine it to run N from the lat. 25°: 00 N. The night leads too many into a voyage to eternity, by depending too much on this falacious information; I could therefore not forbear again hinting at it in this place. Nor can we in the very instance under the present consideration, too much admire the extensive goodness of the All-gracious Ruler of the universe towards us weak mortals, by providing so facile a navigation for the regions of the west, by means of a velocious current, and by so disposing the several shores of this mazy labyrinth of reefs and keys, as to cause this useful current to run in a direction N E, and at so great a rate as 3 or 3½ miles an hour, by which means we are enabled better to avoid the eminent dangers of this reef, when it becomes a lee shore, for the violence of the easterly gales beats the gulph water over the reef, so as to destroy the effects of the flood tides, by causing a constant reverberating current from the shore over the reef, insomuch that a vessel riding under the reef will lay with her stern to windward. I once came out from *Matacombe*, and was scarce clear of the reef before I was overtaken by a gale from the eastward, which was very violent; it was 5 o'clock in the evening, and it being too dark to attempt a re-entrance of the reef, I was forced to heave the vessel too, which I did under her balanced mainsail; she was a heavy schooner of about 70 tons, and a dull sailor: the succeeding night I passed in the deepest distress of mind, seeing the burning of the breakers in constant succession on the reef till past 1 o'clock; the storm continued till 10 next morning, when I made sail to the northward, and at noon to my utter astonishment, I had an observation of the sun's altitude, which proved me to be in lat. 26°: 50' by which I had made a difference of lat. of 118 minutes in the short space of 19 hours (17 of which I lay too). I think this so extraordinary an instance of the rapidity of this current, and so evident a proof of the reason of the increased velocity thereof, that I could not omit relating it. When I treat of the *Behama* bank, it will however appear, that on that side the shore is not so steep, and therefore not so dangerous.

I am an utter enemy to all theoretical systematic positions, which has caused in me an indefatigable thirst for finding in my experience, causes for all extraordinary appearances, be they what they will. And my experimental position of the cause of the increase of the velocity of this current, during the gales that blow contrary to its direction, is no other than the reverberating current, occasioned by the swelling of the water within the reef, which in the memorable gale of October, Anno 1769, when the *Ledbury* was lost, was no less than 30 feet above its ordinary level; which height will appear the more surprizing when we come to consider the spacious surface of the sounds that were filled by it.

Having thus led the careful mariner through the reefs into the open gulph, there only remains for me to give a caution against the *Memory Rock*, for which ample instructions are given in a hint at the very beginning of this tract. My next care will be to give some

INSTRUCTIONS TO PEOPLE BOUND FROM THE EASTWARD OVER THE BAHAMA BANK INTO THE GULPH OF MEXICO.

First take care to make the south end of *Abaco*, commonly called the *Hole in the Rock*, in lat. 26° 4 from thence steer W b S 13 or 14 leagues, which course and distance will carry you to *Stirrups Key*, being the N westernmost of the *Berry Islands*, then run S W till you can but just see it off the deck of a vessel of about 100 tons burthen, you will then be in 8 or 9 fathom water, and by my calculation 7 leagues from it; next hawl up S W b S, and keep no nearer to the southermost *Berry Islands* than in 3 fathom, which will be when you have run on that course 12 or 13 leagues; you will then begin to come on the middle of the bank, where you may depend on finding no where less than 15 foot.

Both on coming on to the bank, and in going off from it, you will find a very strong tide, which sets right on or right off: it is easily observed to go along by the ground, the water being very clear and transparent; on the middle of the bank you will find little or no tide. If the wind hangs too far to the southward, it will follow that you are forced to the westward; in that case make no delay in coming to an anchor, lest you fall in with the bars that lie to the south and westward of the *Bemini's*, and extend near three leagues; get under way as soon as you can steer south; you must see no land after you leave the *Berry Island*, till you make the *Roques*, or *Orange Key;* the *Roques* are four in number, without bush or shrub on them; from the *Roques* to the *Dog Keys*, which are the westermost of the *Double-Headed-Shot*, the course is SSW, and distance about 17 leagues, yet the current will sometimes force you on them in steering SW.

Should you sail for the *Dog Keys* in the night by all means keep clear of them, they and the *Deadman's Keys* are a number of bare rocks, perhaps an hundred, or an hundred and twenty, about the size of a vessel, and some less, but in general there is water plenty between and about them; S b E and 5 leagues from the *Deadman's Keys* is *Key Sal*; here are some sunken rocks; the best way is to run SW from the *Roques*, and not to run the 22 leagues before morning; if in the morning you find yourself in ocean water, run S W b S for the *Matancas*, if not keep down west till you be in ocean water, then hawl up for the cost of *Cuba*, & run down it till you are abreast of the *Bay of Honda*, from whence you must steer over NW, or thereabouts, which will carry you clear of every thing into the Gulph of Mexico.

DIRECTIONS FOR GOING TO THE MISSISIPPI.

If you are bound into this river keep the NW course till you are in lat. 29° 10 N, then run down for the river on a W course; if you happen to fall in to the northward of the mouth, come no nearer than 15 fathom, when you will have coarse brown sand; then run S, or S half W, keeping your lead going, till you come on soundings so soft that the lead will bring none of the mud up, unless it be woolded with canvass; if then the wind be free, run the above course or rather more westerly; but by all means take care you are not to the southward of the last mentioned latitude when you run down on a W course, for fear of falling in to the south of the river: When you approach the rivers mouth, which you will by running about 4 miles on the above soft soundings, you will see the color of the water alter, and it will appear like a shoal; this is occasioned by the current of the river mixing with the sea; but you need apprehend no danger, for there is from 25 to 30 fathom water, therefore run boldly in till you have about 8 fathom, the bank is pretty steep; when you are in this depth, you will see some mud islands about as large as a vessel of 150 or 200 tons; from among which you will perceive the river's mouth; the opening between these islands is about 100 fathoms wide: there are some Spanish pilots kept at *Fort Balize*, who give a very indifferent attendance; but if you see a launch coming out, you may depend on its being them, and your best way is to run directly for them, keeping off and on when you think yourself too near.

If no launch come out, and the weather fine, come to an anchor, there is little or no danger, and if need the current will always carry you out again, provided you keep opposite the channel.

In going up the river it is necessary to keep within about 20 feet of the land, or of the drift wood that lies along it, except when you find it lodged on points or banks, in such place give it a reasonable birth.

Observe also where you see the timber of a dwarfy and shrubby growth, or where willows grow, on the points in such places the water is shoal, and it is requisite to give the land a birth of about 100 feet.

There are however no shoals of any consequence, till you are about 10 miles past the *Detour aur Anglois*, (*Anglice* English reach), here one Mr. *McCarty* has got a plantation on the larboard side, where is a point called *McCarty's* point; from this a shoal runs off near one third of the way over.

It is almost needless to mention that in this (as in all other rivers) vessels ought to keep on the lee side, for on the other side they would be becalmed.

Unless in a case of the utmost necessity, (such as fears of seizure at *Orleans* or otherways) let not go an anchor, for it's a thousand to one but you will loose it if you do; the appearance of the shore will however tell you whether the logs are numerous or not at the bottom: a buoy will not watch.

DIRECTIONS FOR VESSELS WHO ARE BOUND TO PENSACOLA, AND HAVE RUN DOWN BY JAMAICA, OR ON THE SOUTH SIDE OF CUBA.

After making *Cape Anthony*, your course for *Pensacola*, is N W b W ½ W, the distance is 178 leagues, go nothing to the westward of this course, and if there be no currents to deceive you, it will carry you about in with the middle of St. *Rosa Island;* currents are here frequent, and they often change suddenly; in which case you may be carried many leagues either eastward or westward of the cape.

When you come as far North as the lat. 24° 00 N, keep a good look out for the *Dry Tortugas*, in order to avoid that danger, if you see them not, till you pass the lat. 24° 4 N, I would advise you to keep as far up as NNE or even NE b N if the wind allows, till you get soundings; and depend on it, if you are to the eastward you will strike soundings in about 40 fathom, in lat. 25° 00 N; from hence you may steer NNW: it is a sure sign, that the sooner you strike soundings, the more easting you have made; and the longer you run without soundings, the further you are westward.

If you fall in as far eastward as *Cape St. Blas,* you will strike soundings above the lat. 29° 00 or 29° 5 N, your first soundings will be in about 90 fathom, on a muddy or oozy bottom, which is the same everywhere on the edge of the bank; the bank being pretty steep, you will soon be in 60 and 40 fathom, coarse blackish sand, and a few shells in spots: further in your soundings will decrease rather irregularly, on account of some knolls; one of these which lies about 3 leagues south of the cape, has only 18 feet water on it, the depth all around it is about 4 fathom; the pitch of the cape shoal lies in lat. 29° 38 N, about 5 miles out from the land, and the extreme depth of water on it is only 14 or 15 feet: the soundings continue however to the shoal pretty regularly as you go in, till about 4½ or 5 leagues from the pitch of the cape; therefore if you come in upon this part of the land during night, you may run in boldly upon 11 fathom, a hard sandy bottom with broken shells, all the knolls lie within this, they are however only to be dreaded by large ships; this bank is most plentifully stored with fish, especially dolphin for the tow-line sports; the pitch of the cape is known by the appearance of a gap in the land about 1½ or 2 leagues to the eastward of it, in which gap stands a very large single live oak tree; about 4 leagues to the NW of the cape is a middling good roadstead, where in case of easterly winds it is safe riding in 4 or 5 fathom, black mud and shells; and about 3 leagues further north is the bar of *St. Joseph,* the entrance into which harbour will be hereafter described.

If you fall in still more to the eastward, you will find the water clearer, and even in 12 or 13 fathom, it is of a dark blue like the ocean; the soundings begin in 28° 30, or even sooner, and the bottom is fine sand, mixed with coral, shells, and some spunges.

In case you fall in with the coast of *St. Andrews,* you will not strike soundings before you be in lat. 29° 15, or 29° 20 N, same ground as off the cape; but as you advance towards the land you will in many places meet with coarse, muddy grey sand, mixed with black specks, and at some casts the lead will bring up fragments of coral; about the lat. 29.30 N you will find 18 fathom, sand with small shells; and you may then just see the land of *St. Andrews*, to the westward of the inlet; which land has a woody flat appearance, with an even white beach, and a bold shore, so as to have 10 or 11 fathom water within a mile, or a mile and a half from the strand; the coast trenches nearly W by N and E b S.

But if you fall in so far to the eastward as to be right off the entrance of the bay of *St. Andrews*, or between that and *St. Joseph*, the coast is not quite so bold, and the depth of 10 or 11 fathom is but just in sight of land, the bottom is sand and small shells.

The entrance of *St. Andrews* bay lies in lat. 29° 49 N, and admits only of small vessels.

If your landfall be any where between the coast and bay of *St. Andrews*, and the bay and island of *St. Rosa*, you will strike soundings from lat. 29° 20 to lat. 29° 45; the farther north you run without sounding the farther westward you are; the land here stretches E half N and W half S, all these are marks by which the mariner may know the true place of his vessel on this coast, which is too level to distinguish particulars thereon at a distance: if you strike no soundings till lat. 29° 55 it is a sure mark that you are abreast of some part of *St. Rosa* island, which is easily known when you come near, being no more than a long and narrow slip or sand-hills, with here and there some groves of pine trees scattered on it; towards the west end the beech is exceeding white, and some of the sand-hills loom like lofty white buildings, or vessels under sail, especially when not too near; the bottom here is a white sand, with here and there a spot of coral, it shoals very gradually; within the island is a sound which is from 1 to 3 miles wide, narrowest at the east end, so that if you fall in towards that end it is difficult to know it to be an island, though you are within 3 miles of it; the shore is so bold that you may run down along by it at 1½ or two miles off, where there is generally 10 or 11 fathom; to the eastward you can only see the water over the land in spots, though are you at masthead.

From *Cape Blas* to *St. Rosa* bay the course and distance is about NW b W 27 leagues; from that to *Pensacola* bay is W half S between 16 and 17 leagues; and from that to *Mobile* bar W between 14 and 15 leagues: *Cape St. Blas* being so surrounded with rocks and shoals, I would advise by all means to avoid falling in with it, a large ship ought by no means to come nearer than within 6 leagues of it.

For the better knowing of this coast, I shall make some further remarks, because the coast's running so nearly E & W, and being every where so nearly alike in its level & woody appearance; the most skilful and experienced pilot may here be deceived, especially as the advantage of observations, which is so great a help on a N & S coast, fail here totally.

In coming on this coast, if your land-fall be off the island *St. Rosa*, your first soundings will be about 80 fathom (little more or less) oozy ground, this depth and bottom is about 15 or 16 leagues from the land.

If you fall in about 10 leagues eastward of *St. Rosa* bay, you will see the beech very white, and no sand hills on it, which last circumstance distinguishes it from the land further westward; near the beech is very little growth, besides shrubs, and brushy plants; if you stand pretty well in, you may, from the masthead see that an extensive *Savannah* (or plain) is situate within this shrubbery; which is an other mark to distinguish it from the land to the westward, which being an island, water is seen over it.

The bar of *St. Rosa* only admits craft of 6 foot draught. As you come near this inlet or the east end of *St. Rosa* island, it is difficult to know that it is an island; the woods grow close to the beach, which is likewise very white, and full of sand-hills, these being contrasted by the dark green of the bushes, are of as dazzling a white as snow: the island as before observed is a narrow slip of sand-hills, with a very few trees scattered over it, mostly towards the west-end, and from masthead seems about a cables length over, within appears the sound; the wood-land on the north shore of it is of a middling stout growth, and the trees stand pretty close to the water-side; having run down about two thirds of the length of this island, (in other words) about 12 leagues or upwards, you will see several of those remarkable sand-hills above mentioned which appear at a distance like buildings or vessels; approaching the bar of *Pensacola* still nearer, you will see a pretty high bluff point of a redish colour; about 3 or 3½ miles north of the island, on the main land, this point is called *Deer-Point;* when you come abreast of this, you will see the vessels (if any there are) riding at anchor before the town of *Pensacola*, and a watch-house (nick-named a fort), on St. *Rosa Island*. If a guard is kept here, which is commonly the case, you will see a flag hoisted on it if your vessel is a topsail vessel, or a pendant, if she be boomsailed, and a gun fired; which signal is made on account of your approach: if the weather is good, (and the attendance a little better than it was during the time of my residence there) a canoe or barge will come off with a pilot; if none comes off, you may freely venture with a leading wind to run in by my plans. Off of the Look-out on St. *Rosa Island* lies a spit, which you must avoid by not bringing the watch house more northerly than N N W, till you run off in 5 or 6 fathom water, then by all means keep that depth, until you bring the middlemost or highest red cliff, which opens with the west end of St. *Rosa* island, to bear N ¼ E by the compass, and steer directly for it; this course will carry you over the bar in 20 or 21 foot water. The tops of these cliffs are built full of barracks, block-houses, and other military works; so as to appear like a small town. When you are over the bar steer N b W, or N N W, to clear a shoal that stretches near two thirds over from the west end of St. *Rosa* island, on which is 10 foot water, hence it is called the 10 foot bank. The lat. of this bar is 30°: 19' N.

If you fall in to the westward of *Pensacola*, on the *Mobilian* coast, you will get no soundings till you are near or in lat. 30°: N, the edge of the bank is 80 or 90 fathom deep, on a soft muddy and oozy bottom; but it is steep and shoals suddenly, and by the time you run about 2 leagues, you will not have more than 30 or 40 fathom on a soft bottom of very fine white or grey sand, mixed with mud and shells, and you will be in 8 or 9 fathom when 3 or 4 leagues from the land. The coast runs due E and W; large trees come close down to the water side; the beach is not so white; the sand-hills few, and neither water nor savannah to be seen within land; all which circumstances sufficiently distinguish it from the eastern coast. Observe also that the further you go westward the softer the bottom is.

West of *Mobile* the coast is lined with islands, as represented in the map; the bank is steeper, the ground softer, and the land in sight almost as soon as you strike bottom, at least when you are not above 5 or 6 miles on upon soundings. The principal harbour here is between *Cat Island* and *Ship-Island*. The whole of the navigation in and about those islands, through the *Rigolets,* and by the lakes to *Manchac,* is so plainly laid down in my map, that any wordy elucidation of so minute a matter, chiefly for the use of boats, would be prolixity in me.

I will therefore finish this account of the coast of *West-Florida,* by informing my reader that the *Chandeleur* or *Candlemas* islands, were formerly only two islands; but the hurricane of 1772 has cut them into so many parts, that they lie, or at least did lie in near the same number and form as in the map is represented, the channels between them are deep, but barr'd at each end. Just within the north end of these islands, about 5 miles north of the *Free-Masons* islands is a good road, as there is also between *Breton* island and the back of the *Grand Gozier*, a land of islands about 12 miles S S W from the southernmost of the *Candlemas* islands; *Breton* island lies about 3 leagues W from *Grand Gozier*, and this last about 8 or 9 leagues north of the entrance of *Missisippi* river. The depth of water in both these roads is from 20 to 24 feet, and the bottom a fine sand.

If you are bound out of the *Missisippi* eastward, to go through the gulph of *Florida*, you ought to endeavour to make the *Tortugas*, in order for this steer E S E from the *Balize*, run that course till you come on the edge of soundings, between lat. 26° and 27° N, by this means you will not be plagued with the trade wind, in the way of which you would fall too soon, before you got your easting. I remember a vessel bound from *Pensacola* to *Carolina*, who was drove as far to leeward as Cape *Catoche*, and after being a long while (I think 4 or 5 weeks) out, was obliged to return to recruit her stock of provisions; this was doubtless occasioned by not getting her easting while she was out of the way of the *trade*. After you are on this edge of soundings, direct your course more southward, you will find very regular gradations of the depth, such as my map lays them down, you may depend on them. It frequently happens that vessels fall too far to the westward, as was the case with one from *North-Carolina* some four or five years ago, who was either lost or taken near the bay of St. *Bernard* and the people underwent numberless hardships, as well among the Spaniards as savages; I will for the sake of such who will at any rate avoid this danger set down some

INSTRUCTIONS TO GO TO PENSACOLA, ON A ROUT DIFFERING FROM THE FORMER.

The *Dry Tortugas* lie in lat. 24°: 25' and stretch northward as far as lat. 24°: 43' N. the south end lies N 40°: W. 31 leagues from the *Havana*, or N ½ W 22 leagues from *Bahia Honda*, the direct course from the *Tortugas* to *Pensacola* is N 34° W, and the distance 142 leagues. But the safest way is to run N ½ E 35 leagues, by which means you will make the land in lat. 26° 46 N; where is a large harbour called *Charlotte* harbour; here, in case of necessity, you may refresh, as it affords excellent water in many places, especially on a high island, whose north end is a broken bluff, and which shews itself very remarkable as soon as you are well shot in; there is likewise plenty of fish, and the islands are stocked with large herds of deer; there are 4 or 5 inlets into this bay; but the one that lies in the above latitude is deepest, it has 15 or 16 feet water on its bar; the southernmost is the next best, and has 14 feet on its bar; this lies in lat. 26° 30, and is remarkable for the coast taking a sudden turn from N N W, to directly west, only for about 9 or 10 miles; when it again resumes its former direction: this nook in the land, forms what the Spaniards call *Ensenada de Carlos*, i.e. *Charles's-Bay*, the piece of coast that trends E and W, is the beach of an island called *Sanybel*, this place is further remarkable for a great number of pine-trees without tops standing at the bottom of the bay, there is no place like to it, in the whole extent of this coast; the northernmost entrance is likewise remarkable for a singular hommock of pine-trees, or a grove standing very near the beach, than which there is none like it any where hereabouts; the course and distance from this place to *Pensacola* is N 49° W 109 leagues; N 36° W about 20 leagues from this place is the bay of *Tampe* or *Spirito Santo*, from whence to *Pensacola* the course and distance is N 52 W 80 leagues, either of these courses will steer clear of Cape *St. Blas* shoals, and you may in case of currents humour your course, so as to have easting enough.

The navigation in and through the bay of *Juan Ponce de Leon*, to *Punta* Largo, or *Cape Roman*, and as far as *Charlotte* harbour, being fit only for turtlers, fishermen, and other small craft, I will not say much about it; inspection of my charts where that part of the coast is very faithfully laid down, will suffice such small fry.

Juan Ponce de Leon, was one of the first discoverers and explorers of *Florida* and some remarkable transactions between him and the *Coloosa* savages at this place, have given occasion to the bay being called after that adventurer. But our wise map-makers, from the compilers of the quarter waggoners, down to the sagacious *William Gerard de Brahm*, Esq; have corrupted it into *Ponio* bay; tho' the latter has not forgot to change it into *Chatham-bay;* but what connexion the earl or the fort of that name have, or had with this place, is to me a secret; Mr. *de Brahm* does little honor to either, in calling this flat after them.

The fort at *Apalache* being deserted, the coast very flat, and the bay inconsiderable; I shall omit saying any thing about it, as my maps shew with sufficient accuracy what may be expected and done by such small craft, as may have occasion to call there.

Having already said what was most wanted of the passage over the *Bahama* banks, and having nothing material to say concerning the old streights of *Bahama:* I shall end this general account of the coast, with some further

REMARKS CONCERNING NEW PROVIDENCE, AND THE BAHAMA BANKS

This island is (by some mistake) laid down in my map, about 5 or 6 miles further northward than it ought to have been; my observation of lat. in the harbour was lat. 25°: 4′ N. and I rekon its long. about 77°: W from London.

From *Providence* to the S Eastermost key of the *Berry Islands* the course and distance is about N W 11 leagues.

From this key to the entrance on upon the bank between the *Blackwood Bush* and *Joulter Keys*, at the North end of *Andros* island, is about W 8 or 9 leagues: as the bank is bound with a reef here, you must pick your way through that, which you may, as there are several swashes, which though they are narrow, have no less than 11 or 12 feet thro'. The water being very clear in this part of the world, this picking ones way through a shoal, is attended with little or no difficulty.

When you first come on upon the bank, you will see some scattered heads of rocks and spunges; but there being no danger, except what is very visible, I need only tell, that by running W S W about 12 or 13 leagues you will come out 1½ leagues to the southward of the *Riding Rocks*, on the west part of the bank; from whence by inspection of the charts, and some attention to former remarks, you may easily find your way either to *Cuba*, or the *Florida* shore.

N.B. In coming from the *Florida* shore this way, by the *Riding Rocks*, you ought to endeavor the making of the South Eastermost *Berry Island*, early in the morning, which precaution will give you a great benefit with regard to safety in your run to *Providence*. I need not tell how great the danger is of coming among shoals and broken land during night.

I would say something with regard to the passage by *Bemini*, but it being the shoalest, and the bars of the Eastward of *Bemini* making it very dangerous, I think no new comer ought to go there, without a pilot; I shall therefore say nothing of it.

There are several small harbours on the N E part of the *Berry* where water and other refreshment may be had; but as they are seldom frequented, but by the people of *Providence*, it may be superfluous to say any thing about them; the maps shew their situation.

From *Providence* to the *Hole in the rock*, South end of *Abaco*, is N about 20 leagues.

The North, or *Grand Bahama* bank, is little frequented but by whalers and turtlers; and on account of its iron bound reefs is dangerous to approach.

The passage from the *Hole in the Rock*, towards the *Gulph of Florida*, is already explained; yet I will here add, that it is necessary to give the West end of the *Grand Bahama Island a* birth; both on account of its shoals, and if the wind should hang Southwestward you might be imbayed.

I should conclude this part of the work, with some directions for the entrances of particular places; and begin with

DIRECTIONS FOR THE HARBOUR OF SPIRITO SANTO, OR TAMPA BAY

This harbour which is very capacious, will admit large ships and is extremely well calculated for a place to refresh at; here is abundance of wood, water, fish, oysters, clams, venison, turkies, large and small water-fowl, &. The harbour is made by a range of islands lying before it; the southernmost of which is called *Long-Island:* its North end is called *Grant's* point, in honor of governor *Grant;* the next lying about a mile North of it, is called *Pollux Key;* another about 1¼ mile to the N W b N of that is named *Castor Key*, in honor of two privateers, one of which was commanded by the late Capt. *Braddock* of *Georgia:* these two vessels cruized in those seas about the year 1744 or 1745; and Capt. *Braddock* was generally acknowledged the first Englishman who explored this bay. I have seen his original draught which (considering the circumstances under which it was taken) was pretty exact. Next is a cluster of keys called Mullet keys, lying between 2 and 3 miles E b N ¼ N from the north end of *Castor* key; a shoal runs off from each of these to the westward, to that which runs off from the *Mullet* keys the *Spaniards* give the name of *Restingo Largo*. On the *Mullet* keys are huts built by the *Spaniards*, who resort here for the purpose of fishing. *Grant's* point lies in lat 27, 44, and the south end of the *Mullet* keys in lat. 27°: 48 N.

The coast, as has been already observed, has pretty regular soundings, but none very deep; to enter this bay by either of its inlets, observe the following directions, and they can hardly fail of carrying you safe.

The land is low and not visible till you are within about 8 miles from it, where you will have 6 or 6½ fathom water; the chief growth on the keys are mangrove and blackwood bushes.

To run in by *Grant's* point, bring that to bear N E ¼ E, then run in on that course till the south end of *Pollux* key bears N E ½ N, when you will be on the bar of this inlet, where you will find 16 foot water; the bar is short, you must run in on the same course till you are nearly abreast *Pollux* key, and you will have 3, 4, and 5 fathoms; when you be almost abreast of the key, steer E, and you may run in as the draught shews.

To run in between *Castor* and *Pollux* keys, keep in about 5 fathom, till you bring the north end of *Pollux* to bear about E b N ¼ E, then run in on that course till the south end of *Castor* bears N E b E ½ E, and you will presently be in about 17 feet, then steer about E N E directly for the midway between the two keys, and you will continue for about ¾ of a mile in 16 or 17 foot, shoalest under the north bank; when you are over this you will have 3½ or 4 fathom, and by keeping a little to the northward of your last course after you have cleared the keys, you may run up the bay without scruple.

To go in between *Mullet* keys and *Castor* key, which is the principal inlet, and by the *Spaniards* called *Boca Grande*, you must run in about 5 fathoms, till you bring the south part of the *Mullet* keys to bear E ½ S, then sail on that course till the north end of *Castor* key bears E b S ½ S, and you will find 22 or 23 foot water, steer that course till you deepen your water to 6 fathom, then run E inclining rather to the south shore, if any; as soon as you have doubled

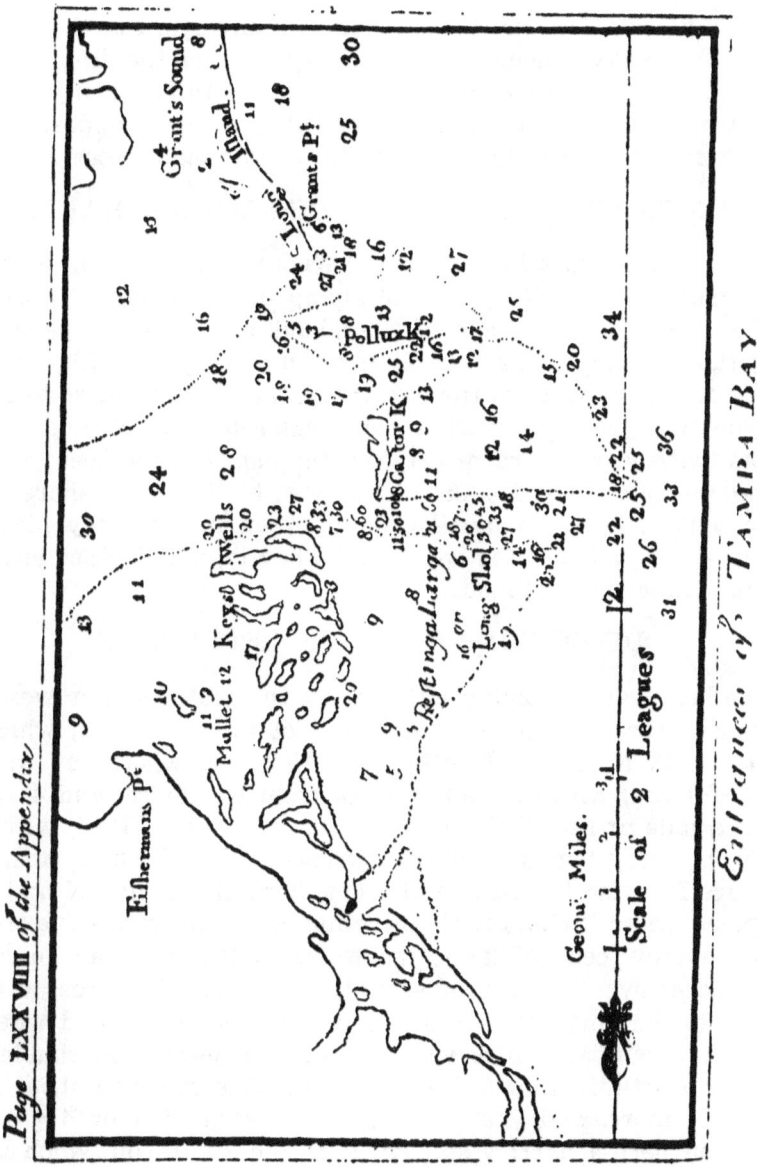

Castor key you may anchor under it, or run up by the *Mullet* keys, or farther in, as the draught points out.

Small vessels need not be so scrupulous in regard to these marks, the banks or shoals themselves are pretty deep, as the draughts point out. A.D. 1769, I was employed above 6 weeks in surveying this bay, and after sinking my boat in *Manatee* river, where I suppose she lays now, I went across the *Peninsula* to St. *Augustine* on foot.

DIRECTIONS TO SAIL INTO CHARLOTTE HARBOUR

The northernmost inlet to this harbour is in lat. 26, 46 N; the surest mark to find it by, is the clump or hommock of pine trees standing near the north end of the island, as mentioned page lxxvi; bring those trees to bear E S E, and run in for them till you are in 15 foot of water, which is the bar, then change your course to E N E, and you will presently deepen your water to 4, 5, 6, and even 10 fathom, with working room between the banks for a fleet; keep the north-breaker on board, and run in close to the north shore, which is the south end of *Gasparilla* island, when in, you may pick your anchoring ground; but the bay is flat when in, and further you go in the less water you will find.

DIRECTIONS FOR ST. JOSEPH'S BAY

The best way in coming either from the southward or westward, is to make the coast to the northward of Cape St. *Blas*, which lies in lat. 29, 42 N, about 10 miles from the cape, where you will find 6 or 7 fathom, within about 2 miles from the shore, and 4 fathom within a mile or less. This coast is a narrow slip of land, with some bushes and very few trees on it, it trenches N b W and S b E, up to the place I advise to make, and from there almost due N and S. At this place where it changes its direction, are two remarkable trees on a very narrow neck of land; the water in the bay may be seen in many places over this slip of strand. From these two trees you may coast it to the northward within a mile of the shore; and when you have run about two miles you will see two more remarkable trees standing a little further in land; you will then continue about a mile or a little more, and you will have the depth of 4 or 4½ fathom water, but here a narrow spit runs off for a mile, on which is only 12 foot, therefore keep your lead going, and when you once lessen your depth to 3½ fathom, draw off, steering N W or W N W till

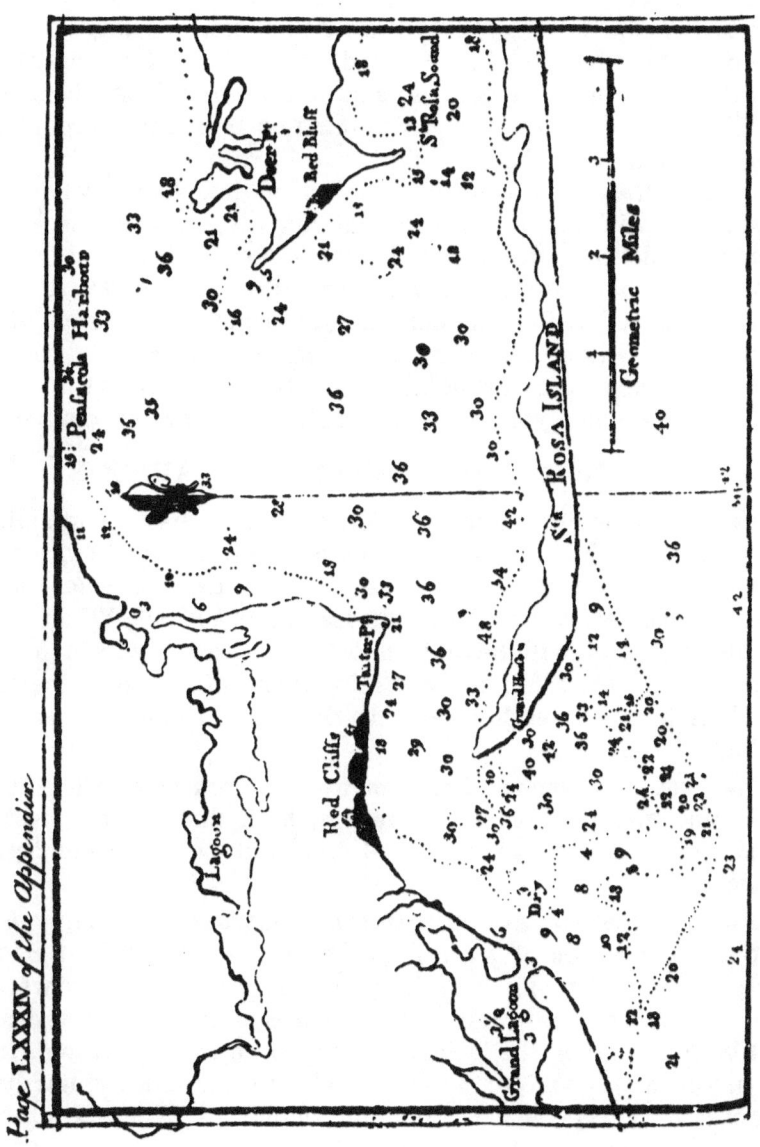

you deepen again to 5 fathom; you will then open a large red bluff on the main land in the N E quarter, bring the east end of it to bear N E b N, and then run on that course till you deepen your water to 6 fathom, and you will see two trees, one large the other small, close together; to the eastward of the bluff steer eastward till you bring these on with the east end of the bluff, keep them so and you will be in the best channel. In running on this course about 2 miles or 2½, you will be on the bar, which has 17 foot water on it in spring tides at low water; as soon as you are on this bar hawl round to the eastward with an easy sweep, and you will presently be in 5 or 6 fathom: by degrees draw round to the southward, and you may anchor any where in that depth, near the west shore. To know exactly when you are on the bar, take notice of four clumps of trees on the eastern shore, and bring the third (counting from the northward) as marked in my draught, to bear S4OE, in one with the point of St. *Joseph's*, when you have this mark on, you are on the bar.

DIRECTIONS FOR PENSACOLA HARBOUR

Having, as before directed, brought the highest or middlemost red cliff to bear N¼ E, you will see a clump of trees (in land) on with that part of the cliff; this is the best mark to lead you over the bar in the best water, being no less than 20 foot. When you are over the bar, you will have from 4 to 6 fathom within a cable's length and an half off the western breakers; these shew themselves very plain in fine weather; and at their north end is a small dry sand bar.

When you are over the bar, run west till you bring the aforesaid trees in one with the hollow between the high cliff and the west cliff, by this means you will avoid the 10 foot bank before taken notice of on page LXIII.

To know when you are up with this bank, observe when you open the straggling woods of *Deer Point*, with the west end of St. *Rosa* island; when this mark is well open, you may hawl eastward for *Tartar Point*. Take care not to approach the starboard shore nearer than 3½ fathom, nor the larboard shore more than 4 fathom. In mid channel you will have 5 or 6 fathom. The bottom is a regular hollow, therefore you may depend on your soundings shoaling gradually down to the above depths on each side.

In working up after you are within the 10 foot bank, keep the

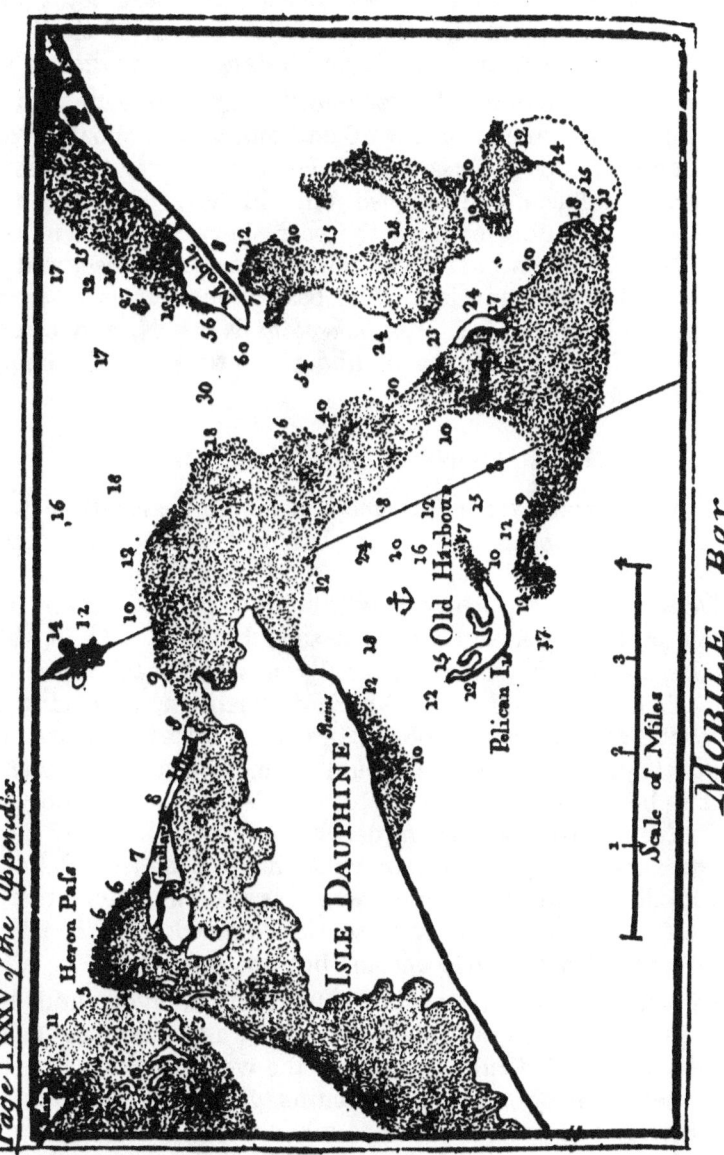

block-house, or fort, open with St. *Rosa Point,* by this means you are in no danger from the ten feet bank, and in standing northward keep *Deer-point* open with *Tartar-point;* off this last point you will have three fathom within less than a cable's length of the beech, in standing for St. *Rosa island,* you have no danger but what is visible.

After you have doubled *Tartar-point,* your course up to town is NNE ½ E; mid channel you have 6 fathom, and it shoals gradually down to 4 on each side; in beating up, come no nearer the west shore than to bring a large white-house (the highest in town) to bear N N E, and in standing over to the east shore, bring it no further west than N ½ W; this house was the government house, and stands at the east end of the garrison. The best anchorage is to bring the above house to bear N ½ E, *English-point* NE b N, and *Deer-point* SE b E ½ E; at the distance of about a mile from the shore, the depth is about 4 fathom.

DIRECTIONS FOR MOBILE BAR.

In going westward from *Pensacola,* come no nearer than within 5 miles from *Mobile-point,* and when that bears N, run till you bring the east end of the *Sand-key* to bear N 27°: W in one with the east end of *Dauphin-island;* then you will be on the bar in 13 or 14 feet, run up on said course, and you will soon have 18, 24, 30, 40 feet; the bar is short: continue to go so till you are abreast & pretty near the *Sand-Key,* then haul up N ¼ E, and run in, observing not to borrow too nigh on the starboard breakers; after you are abreast *Mobile-point,* you may gradually haul up more eastward; then bring the point to bear about SW, or SWbS 2 or 2½ miles off, and you may anchor in from 3 to 4½ fathom depth. Or you may bring the south end of the sand key just shut in with the S E end of *Pelican* key, and run in till you are in 15 foot, which is the bar; then steer W, or W½ N, till you bring the first-mentioned marks on, & run in as before directed: This way is deeper, but not so certain.

There are two swatches thro' the east breaker, and the old harbour (which according to *Charlevoix,* was ruined by an earthquake) lies back of *Pelican Island,* and the west breaker; the draught explains them. On shore, are the ruins of the *French settlement,* which formerly was on that island; what is left most perfect, is the remains of a guard-house, in the sand-hills, and a remnant of a salt-work on the beach, all the rest has been swallowed by the sea; after great storms, it is not uncommon to find numbers of human bones at this place. Being got into this bay, the way up to town is easy, but the whole of the way is shallow, no where exceeding 16 feet; there are no shallows, except between Point *Claire* and *Roebuck* river; where is an oyster-bank, which is easily avoided by keeping nearest to the larboard or west shore.

Having been pretty explicit about the tides on the eastern coast, it may naturally be expected that I should say as much about the tides in the *Gulph* of *Mexico;* but little or nothing can be said about so unstable a matter, from *Sandy-Key* and *Sandy-Point*, or *Punta Tancha,* through the whole bay or *Juan Ponce de Leon,* up as far as *Punta Largo* or *Cape-Roman,* it runs tide and half tide in the same manner as at *Plymouth,* the *Needles* and *Wight,* in *England;* that is to say, three hours flood, then three hours ebb, next nine hours flood, and lastly nine hours ebb, it does not rise to an equal height in all places, nor does it run equally rapid in every part; in some narrow places, I have met it a mere fall, and in almost every gut among the many islands in this bay, I have found it as much as four stout men could do, to stem the current in a *Moses.* From *Cape-Roman* northward and westward in every part of the *Mexican Sea,* the tide seems to ebb and flow once in 24 hours; but it being very much governed by the winds, this circumstance happens very irregular; for with a S or SW and W wind, it will flow much longer than it will fall; whereas with a N or N E and E wind, it will ebb much longer than it will rise; consequently, it happens frequently that at the time of springs, we find less water on a bar than at neap times, and *vice versa.* I never could observe it to rise above two feet any where at the highest times; yet its effect on the current of the rivers, is in dry summers very visible, a great way from the sea; and I have been told that instances are not wanting, when the water of *Missisippi,* was brackish above the town of *Orleans.*

There is a constant current in the open *Gulph,* setting ESE, or thereabouts, at the rate of 1 or 1½ knot per hour, of which all observing people coming through that sea are very soon sensible.

The passage to the eastward of the *Tortugas,* is to be depended upon as laid down in my maps; coming through it from the northward, you will see a rip appear like breakers; but in the rip is 18 or 20 fathom, and the moment a ship gets into this rip, she jumps out of soundings.

A note in page 269 of the appendix, refers to the end; the note was accidentally omitted, and intended to be as follows,—

Besides this reason for standing longer off than in, there is one still greater. Every experienced mariner knows, that a vessel will run a distance towards shore in much less time, then she can run the same distance from it; and that the higher the land is she works under, the quicker she runs in, consequently the slower off; hence almost every one in beating upon a lee-shore, will stand out a longer time than in; but few even among the most experienced know the philosophy of this *phenomenon*, against which they so carefully guard. It is that great law of nature whereby all light bodies must fall on the heavier ones, I mean ATTRACTION. To explain this by an experiment, take any vessel, fill it almost with water, put a cork or chip in it, while that remains in the center, it is attracted from ever side alike, and therefore stands fixed, but no sooner is it thrown out of the center, than it will begin to approach the side, and as it draws near, attraction is increased, till at last the velocity of the chip becomes so great, as to run with a considerable violence against the vessel, where it remains fixed; and if it is an oblong piece in shape of a vessel, the same will happen, as when a ship runs ashore stem on, viz. it will wind broad-side too. The explanation of this affair, I am indebted for to the hon. *John Collins* Esq; of *Newport, Rhode-Island,* first counsellor of the colony, and an experienced commander.

It may not be amiss to conclude this appendix with an answer to a question which has been very often put to me, viz. Why are books intended to elucidate navigation on any coast called *Waggoners?* What connection is there between a waggoner and any thing that concerns the sea, or a ship?

In answer to this, I say from the best authority, that the first book of the kind ever published, called the Mirror of Navigation, was wrote and printed at *Enkhuyzen,* the author's name was *Lucas Jansen Wagenaar;* and his book remaining a very long while the only one of that sort, it was usually enquired for by that name, and in course of usage the name fixed on all that followed it.

FINIS

ADVERTISEMENT

The map of the country of the savage nations, intended to be put, facing page 72, was engraved by a Gentleman who resides in the country 60 or 70 miles from NEW-YORK, to whom the plate was sent; but when it was sent back, it miscarried, through the carelessness of the waggoner; and though the publication has been delayed some time on that account, it is not yet come to hand; the reader will therefore please to expect said map with the second volume.

At the first planning of this publication, it was intended only to be a single volume, not exceeding 300 pages, *appendix* and all; but at the request of some Gentlemen, my friends, I have subjoined so many articles, that it swelled imperceptibly to about 800 pages, which made it necessary to print it in two volumes; and as some unexpected accidents, especially the want of a copper-plate printer, have occasioned delays; I will therefore, to atone in some measure (for said delays) to those kind Gentlemen who favoured me with their subscriptions for the maps, deliver them the second volume *gratis*, as soon as it is published: It is now in the press.

I return my most sincere thanks for the liberal encouragement that has been afforded me; and hope my work will be deemed deserving it.

BERNARD ROMANS